EMERSON'S THIRD PART.

THE

NORTH AMERICAN ARITHMETIC.

PART THIRD,

FOR

ADVANCED SCHOLARS.

BY FREDERICK EMERSON,

LATE PRINCIPAL IN THE DEPARTMENT OF ARITHMETIC,
BOYLSTON SCHOOL, BOSTON.

BOSTON:

RUSSELL, ODIORNE, & METCALF.

NEW YORK, Collins & Hannay. PHILADELPHIA, Hogan & Thompson. BALTIMORE, David Cushing. HARTFORD, F. J. Huntington. WINDSOR, Ide & Goddard. HALLOWELL, Glazier, Masters, & Co. CINCINNATI, C. P. Barnes. RALEIGH, Turner & Hughes.

1834.

STEREOTYPED BY THOMAS G. WELLS AND CO.
BOSTON.

PREFACE.

THE work now presented, is the last of a series of books, under the general title of THE NORTH AMERICAN ARITH- METIC, and severally denominated *Part First, Part Second,* and *Part Third.*

PART FIRST is a small book, designed for the use of children between five and eight years of age, and suited to the convenience of class-teaching in primary schools.

PART SECOND consists of a course of oral and written exercises united, embracing sufficient theory and practice of arithmetic for all the purposes of common business.

PART THIRD comprises a brief view of the elementary principles of arithmetic, and a full development of its higher operations. Although it is especially prepared to succeed the use of Part Second, it may be conveniently taken up by scholars, whose acquirements in arithmetic are considerably less than the exercises in Part Second are calculated to afford. While preparing this book, I have kept in prominent view, two classes of scholars; viz.— those who are to prosecute a full course of mathematical studies, and those who are to embark in commerce. In attempting to place arithmetic, as a science, before the scholar in that light, which shall prepare him for the proper requirements of college, I have found it convenient to draw a large portion of the examples for illustration and practice, from mercantile transactions; and thus pure and mercantile arithmetic are united. No attention has been spared, to render the mercantile information here presented, correct and adequate. Being convinced, that many of the statements relative to commerce, which appear in books of arithmetic, have been transmitted down from ancient publications, and are now erroneous, I have drawn new data from the counting-room, the insurance office, the custom-house, and the laws of the present times. The article on Foreign Exchange is comparatively extensive, and I hope it will be found to justify the confidence of merchants. Its statements correspond to those of the British 'Universal Cambist,' conformably with our value of foreign coins, as fixed by Act of Congress, in 1834.

Although a knowledge of arithmetic may, in general, be well appreciated as a valuable acquisition, yet the effect produced on intellectual character, by the exercises necessary for acquiring that knowledge, is not always duly considered. In these exercises, the mental effort required in discovering the true relations of the data, tends to strengthen the power of comprehension, and leads to a habit of investigating; the certainty of the processes, and the indisputable correctness of the results, give clearness and activity of thought; and, in the systematic arrangement necessary to be observed in performing solutions, the mind is disciplined to order, and accustomed to that connected view of things, so indispensable to the formation of a sound judgment. These advantages, however, depend on the manner in which the science is taught; and they are gained, or lost, in proportion as the teaching is rational, or superficial.

Arithmetic, more than any other branch of learning, has suffered from the influence of circumstances. Being the vade-mecum of the shop-keeper, it has too often been viewed as the peculiar accomplishment of the accountant, and neglected by the classical student. The popular supposition, that a compendious treatise can be more easily mastered than a copious one, has led to the use of text-books, which are deficient, both in elucidation and exercises. But these evils seem now to be dissipating.—The elements of arithmetic have become a subject of primary instruction; and teachers of higher schools, who have adopted an elevated course of study, are no longer satisfied with books of indifferent character.

It has been my belief, that a treatise on arithmetic might be so constructed, that the learner should find no means of proceeding in the exercises, without mastering the subject in his own mind, as he advances; and, that he should still be enabled to proceed through the entire course, without requiring any instruction from his tutor. Induced by this belief, I commenced preparing The North American Arithmetic about five years since; and the only apology I shall offer, for not earlier presenting its several Parts to the public, is the unwillingness that they should pass from my hands, while I could see opportunity for their improvement.

Boston, October, 1834.   F. EMERSON.

A KEY to this work (for teachers only) is published separately.

ARITHMETIC.

ARTICLE I.

DEFINITIONS OF QUANTITY, NUMBERS, AND ARITHMETIC.

QUANTITY is that property of any thing which may be increased or diminished—it is *magnitude* or *multitude*. It is magnitude when presented in a mass or continuity; as, a quantity of water, a quantity of cloth. It is multitude when presented in the assemblage of several things; as, a quantity of pens, a quantity of hats. The idea of quantity is not, however, confined to visible objects; it has reference to every thing that is susceptible of being more or less.

NUMBERS are the expressions of quantity. Their names are, One, Two, Three, Four, Five, Six, Seven, Eight, Nine, Ten, &c. In quantities of multitude, *One* expresses a UNIT; that is, an entire, single thing; as one pen, one hat. Then each succeeding number expresses one unit more than the next preceding. In quantities of magnitude, a certain known quantity is first assumed as a measure, and considered the unit; as one gallon, one yard. Then each succeeding number expresses a quantity equal to as many times the unit, as the number indicates. Hence, the value of any number depends upon the value of its unity.

When the unit is applied to any particular thing, it is called a *concrete* unit; and numbers consisting of concrete

3*

units are called concrete numbers: for example, one dollar, two dollars.   But when no particular thing is indicated by the unit, it is an *abstract* unit; and hence arise abstract numbers: for example, one and one make two.

Without the use of numbers, we cannot know precisely how much any quantity is, nor make any exact comparison of quantities. And it is by comparison only, that we value all quantities; since an object, viewed by itself, cannot be considered either great or small, much or little; it can be so only in its relation to some other object, that is smaller or greater.

ARITHMETIC treats of numbers: it demonstrates their various properties and relations; and hence it is called the Science of numbers. It also teaches the methods of computing by numbers; and hence it is called the Art of numbering.

II.

NOTATION AND NUMERATION.

NOTATION is the writing of numbers in numerical characters, and NUMERATION is the reading of them.

The method of denoting numbers first practised, was undoubtedly that of representing each unit by a separate mark.   Various abbreviations of this method succeeded; such as the use of a single character to represent *five*, another to represent *ten*, &c.; but no method was found perfectly convenient, until the Arabic FIGURES or DIGITS, and DECIMAL system now in use, were adopted.   These figures are, 0, 1, 2, 3, 4, 5, 6, 7, 8, 9; denoting respectively, nothing, one unit, two units, three units, &c.

To denote numbers higher than 9, recourse is had to a law that assigns superior values to figures, according to the order in which they are placed. viz.   *Any figure placed to the left of another figure, expresses ten times the quantity that it would express if it occupied the place of the latter.* Hence arise a succession of higher orders of units.

As an illustration of the above law, observe the different quantities which are expressed by the figure 1.

When standing alone, or to the right of other figures, 1 represents 1 unit of the first degree or order; when standing in the second place towards the left, thus, 10, it represents 1 ten, which is 1 unit of the second degree; when standing in the third place, thus, 100, it represents 1 hundred, which is 1 unit of the third degree; and so on. The zero or cipher (0) expresses nothing of itself, being employed only to occupy a place.

The units of the second degree, that is, the tens, are denoted and named in succession, 10 ten, 20 twenty, 30 thirty, 40 forty, 50 fifty, 60 sixty, 70 seventy, 80 eighty, 90 ninety. The units of the third degree, that is, the hundreds, are denoted and named, 100 one hundred, 200 two hundred, 300 three hundred, and so on to 900 nine hundred. The numbers between 10 and 20 are denoted and named, 11 eleven, 12 twelve, 13 thirteen, 14 fourteen, 15 fifteen, 16 sixteen, 17 seventeen, 18 eighteen, 19 nineteen. Numbers between all other tens are denoted in like manner, but their names are compounded of the names of their respective units; thus, 21 twenty-one, 22 twenty-two, 23 twenty-three, &c.; 31 thirty-one, 32 thirty-two, &c. &c. This nomenclature, although not very imperfect, might be rendered more consistent, by substituting regular compound names for those now applied to the numbers between 10 and 20. This alteration would give the names, 11 ten-one, 12 ten-two, 13 ten-three, &c.

As the first three places of figures are appropriated to simple units, tens, and hundreds, so every succeeding three places are appropriated to the units, tens, and hundreds of succeeding higher denominations. For illustration, see the following table.

Duodecillions.	Undecillions.	Decillions.	Nonillions.	Octillions.	Septillions.	Sextillions.	Quintillions.	Quadrillions.	Trillions.	Billions.	Millions.	Thousands	Units
460	725	206	194	007	185	039	000	164	396	205	013	008	741

By continuing to adopt a new name for every three degrees of units, the above table may be extended indefinitely. Formerly, the denominations higher than thousands were each made to embrace six degrees of units; taking in, thousands, tens of thousands, and hundreds of thousands. The mode of applying a name to every three degrees, however, is now universal on the continent of Europe, and is becoming so in England and America.

The learner may denote in figures, the following numbers, which are written in words.

Example 1. Four hundred seventy-eight million, two hundred forty-one thousand, and one hundred.

2. Seven million, six hundred ninety-two thousand, and eighty-nine.

3. Nineteen million, twenty thousand, and five.

4. Eight hundred billion.

5. One billion, six hundred forty-four thousand, five hundred and thirteen.

6. One trillion, five hundred thirty-four billion, three million, eighteen thousand, and four.

7. Two hundred billion, sixteen thousand and one.

8. Eleven billion, one million, and sixty.

9. Five trillion, eight billion, four million, nine thousand, and seven.

10. One hundred trillion, twenty billion, three hundred million, two thousand, and four.

11. Thirty-one trillion, five hundred, and sixty.

12. Six quadrillion, two hundred and fourteen trillion.

13. Two hundred forty-nine quadrillion, seventy-five thousand, and twenty-two.

14. Forty-six quintillion, one quadrillion, nineteen billion, seven hundred and eight.

15. Nine hundred sextillion, three hundred twenty-five trillion, two thousand, and fourteen.

INDICATIVE CHARACTERS OR SIGNS.

The sign $+$ (*plus*) between numbers, indicates that they are to be added together; thus, $3+2$ is 5.

The sign $-$ (*minus*) indicates, that the number placed after it, is to be subtracted from the number placed before it; thus, $5-2$ is 3.

The sign $\times$ (*into*) indicates that one number is to be multiplied into another; thus, 4×3 is 12.

The sign $\div$ (*by*) indicates that the number on the left hand is to be divided by the number on the right hand; thus, $12 \div 3$ is 4.

The sign $=$ (*equal to*) indicates that the number before it, is equal to the number after it; for example, $4 + 2 = 6$. $6 - 2 = 4$. $5 \times 3 = 15$. $15 \div 3 = 5$.

III.

ADDITION.

ADDITION is the operation by which two or more numbers are united in one number, called their *sum*. It is the first and most simple operation in arithmetic, effecting the first and most simple combination of quantities.

The primary mode of forming numbers, by joining one unit to another, and, this sum to another, and so on, exhibits the principle of addition. When numbers, which are to be added, consist of units of several degrees, such as tens, hundreds, &c., it is found convenient to add together the units of each degree by themselves; and since ten units of any degree make one unit of the next higher degree, the number of tens in the sum of each degree of units is carried to the next higher degree, and added thereto.

RULE FOR ADDITION. *Write the numbers, units under units, tens under tens, &c. Add each column separately, beginning with the column of units. When the sum of any column is not more than 9, write it under the column: when the sum is more than 9, write only the units' figure under the column, and carry the tens to the next column. Finally, write down the whole sum of the left hand column.*

1. What is the sum of $370 + 90264 + 1470 + 40060$?

2. What is the sum of $4000 + 570 + 99 + 54 + 273 + 69073 + 4000 + 61998 + 752$?

3. What is the sum of 243+5021+7628+927+64
+5823+742+796+ 5009+ 325 +7426+31186 +
987+6954+2748?

4. What is the sum of two thousand and seven, forty-
four million five hundred and sixty-one, one hundred million, six billion twenty-eight thousand and eleven?

IV.

SUBTRACTION.

SUBTRACTION is the operation by which one number
is taken from another.

The number from which another is to be taken is
called the *minuend*, and the number to be taken is called
the *subtrahend*. The number resulting from the operation shows the *remainder* of the minuend, after the
subtrahend has been taken out; it also shows the *difference* between the minuend and subtrahend, or the *excess*
of the former above the latter. The subtrahend and remainder may be considered the two parts into which the
minuend is separated by the operation; and in this view,
subtraction is the opposite of addition, in as much as
addition unites several quantities in one sum, and subtraction separates a quantity into two parts.

Subtraction is performed by taking the units of each
degree in the subtrahend, from those of corresponding
degree in the minuend, and severally denoting the remainders. When the units of any degree in the subtrahend exceed those of the same degree in the minuend,
we mentally join one unit of the next higher degree to
the deficient place in the subtrahend, and consider the
units of the higher degree to be one less than they are
denoted: this process is the reverse of *carrying* in
addition. One other method may be adopted in this
case; viz. Increase both the minuend and subtrahend,
by mentally adding *ten* to the deficient place in the
former, and, *one* to the next higher degree of units in
the latter. This method is justified by the self-evident

trath, that, if two unequal quantities be equally increased, their difference is not thereby altered.

RULE FOR SUBTRACTION. *Write the smaller number under the greater, placing units under units, &c. Begin with the units, and subtract each figure in the lower number from the figure over it. When a figure in the upper number is smaller than the figure under it, consider the upper figure to be 10 more than it is, and the next upper figure on the left hand, to be 1 less than it is.*

PROOF. *Add together the remainder and the smaller number: their sum will be equal to the greater number, if the work be right.*

1. What is the difference between 70240 and 69418?

2. How much is the excess of the number 482724 above the number 194750?

3. Suppose 479021 to be a minuend, and 38456 the subtrahend; how much is the remainder?

4. 905106392 — 904623724 = ?

5. Subtract fifty-one thousand from one hundred billion, eighteen thousand, five hundred and one.

V.

MULTIPLICATION.

MULTIPLICATION is the operation by which a number is produced, equal to as many times one given number, as there are units in another given number. It is an abridged method of finding the sum of several equal quantities, by repeating one of those quantities.

The number to be multiplied or repeated is called the *multiplicand;* it may be viewed as one of several equal quantities, whose sum is to be produced by the operation. The number to multiply by is called the *multiplier;* it indicates how many such quantities as the multiplicand are to be united, or, how many times the multiplicand is to be repeated. The number resulting from the operation is called the *product.*

The multiplicand and multiplier, considered as concurring to form the product, are called *factors* of the product. Either factor may be used as the multiplier of the other; that is, the multiplicand and multiplier may change places, and the product will be still the same. For example, $4 \times 3 = 12$. $3 \times 4 = 12$.

When a product arises from more than two factors, the numbers may be denoted thus, $6 \times 3 \times 5 = 90$; but, in forming the product, a distinct operation is necessary to bring in each factor, after the two first. The numbers, 6, 3, 5, would, therefore, be multiplied into each other thus, $6 \times 3 = 18$; $18 \times 5 = 90$.

Factors may be arranged in any succession whatever, since the mere order in which they are brought into the operation cannot affect their final product. For example, $5 \times 3 \times 4 = 60$. $4 \times 3 \times 5 = 60$. $3 \times 5 \times 4 = 60$.

The products of small numbers may be committed to memory; but when the product of factors consisting of several figures is required, it is necessary to multiply each figure in the multiplicand by each figure in the multiplier, and denote the several products in such order that they shall represent their respective values. When simple units are employed as the multiplier, the product of each figure in the multiplicand is of the same degree as the figure multiplied; that is, units multiplying units give units, units multiplying tens give tens, units multiplying hundreds give hundreds, &c. When tens are employed as the multiplier, the product of each figure in the multiplicand is one degree higher than the figure multiplied; that is, tens multiplying units give tens, tens multiplying tens give hundreds, tens multiplying hundreds give thousands, &c. When hundreds are employed as the multiplier, the product of each figure in the multiplicand is two degrees higher than the figure multiplied; and so on.

RULE FOR MULTIPLICATION. *Write the multiplier under the multiplicand, placing units under units, &c.*

When there is but one figure in the multiplier, begin with the units, multiply each figure in the multiplicand separately, and place each product under the figure in

*the multiplicand from which it arose; observing to carry
the tens to the left as in addition.*

*When there is more than one figure in the multiplier,
multiply by each figure separately, and write its product
in a separate line, placing the right hand figure of each
line under the figure by which you multiply; and finally,
add together the several products. The sum will be
the whole product.*

Abbreviations of the above rule may frequently be
adopted, as follows.

*When there are ciphers standing between other fig-
ures, in the multiplier, they may be disregarded.*

*When ciphers stand on the right of either factor, or
both, they may be disregarded till the multiplication is
performed, and then annexed to the product.*

*When either factor is 10, 100, 1000, &c., merely
place the ciphers in this factor on the right hand of the
other factor, and it becomes the product.*

*When the multiplier is a number that can be produc-
ed by multiplying two smaller numbers together, multiply the multiplicand first by one of the smaller numbers,
and the product thence arising by the other.*

1. Suppose 479265 to be a multiplicand, and 9236
the multiplier; how much is the product?

2. Suppose 26537 to be one factor, and 873643
another; how much is their product?

3. Suppose the numbers 725, 38046 and 91, to be
factors; how much is the product?

4. What is the product of 62392×4003?

5. What is the product of 248000×9400?

6. What is the product of $24 \times 300 \times 13 \times 10002$?

7. Multiply one hundred five million, by one thousand.

For the purpose of determining whether any error has
happened in the process of multiplication, the following
method of trial, which depends on the peculiar property
of the number 9, and which is called *casting out the
nines,* may be practised.

Add together the figures of the product, horizontally,

rejecting or dropping the number 9 as often as the sum amounts to that number, and proceeding with the excess, and finally denote the last excess. Perform the same operation upon each of the factors; then multiply together the excesses of the factors, and cast out the nines from their product. If the excess of this smaller product be equal to the excess of the larger product first found, the work may be supposed to be right. It is, however, to be observed, that, although this test furnishes satisfactory evidence of the correctness of an operation, it is not an infallible proof; for, if a product chance to contain an error of just 9 units of any degree, the excess of its horizontal sum is not thereby altered.

In order to perceive why the excess above nines found in the horizontal sum of a product, must be equal to the excess found in the product of the excesses of the factors, observe that, by the law of notation, a figure is increased nine times its value by its removal one place to the left; and hence, however far a figure is removed from the place of units, when its nines are excluded, its remainder can be only itself. Therefore, any number, and the horizontal sum of its figures, must have equal remainders when their nines are excluded. This being understood, observe that, since factors composed of entire nines will give a product consisting of entire nines, it follows, that any excess above nines in a product, must arise from an excess above nines in the factors. Therefore, the product of the excesses of the factors, must contain the same excess that is contained in the product of the whole factors.

VI.

DIVISION.

Division is the operation by which we find how many times one number is contained in another. It is the converse of multiplication; the product and one factor being given, and the other factor resulting from the operation.

The number which corresponds to the product in multiplication, is the number to be divided, and is called the *dividend*. The given factor is the number to divide by, and is called the *divisor*. The factor to be found, that is, the number which shows how many times the dividend contains the divisor, is called the *quotient*.

As multiplication has been shown to proceed from addition, so division may be shown to proceed from subtraction. If we repeatedly subtract the divisor from the dividend till the latter is exhausted, the number of subtractions performed will answer to the number of units in the quotient. For example, if the dividend be 24, and the divisor 6, the quotient may be found by subtraction thus, $24-6=18$, $18-6=12$, $12-6=6$, $6-6=0$. Here 6 is subtracted four times from 24, and there is nothing remains; therefore, 4 is the number of times that 6 is contained in 24. In division, this operation is denoted thus, $24 \div 6 = 4$; or thus, $\frac{24}{6} = 4$.

Division not only investigates the number of times the dividend contains the divisor, but it also serves to divide the dividend into as many equal parts as the divisor contains units; the quotient being one of these parts. This effect of the operation may be understood by considering, that, since the divisor and quotient are factors of the dividend, they must each indicate how many of the other the dividend contains.

It may be observed, that all the preceding operations begin at the place of simple units; division, however, must begin at the highest degree of units; for, the number of times that the divisor is contained in the higher units of the dividend must be taken out first, in order that any remainder, or excess above an exact number of times, may be carried down to the lower degrees of units, and divided therewith.

When the divisor is not contained an exact number of times in the dividend, there will be a *remainder* at the end of the operation. This remainder, being a part of the dividend, is to be divided; but its quotient will be smaller than a unit, since a quantity in the dividend just equal to the divisor, gives only a unit in the quotient.

Quantities smaller than a unit, that is, parts of a unit, are called FRACTIONS. Such quantities are commonly expressed by two numbers, placed one above the other with a line between them, thus $\frac{1}{2}$. The lower number, called the *denominator*, shows how many equal parts the unit is divided into; and the upper number, called the *numerator*, shows how many of the equal parts are embraced in the fraction. When the unit is divided into two equal parts, the parts are called *halves*; when divided into three equal parts, the parts are called *thirds*; when divided into four equal parts, the parts are called *fourths*; and so on; the number of the denominator giving the name. For example, if the unit be divided into five equal parts, one of the parts is denoted thus, $\frac{1}{5}$, and called one-fifth; two of the parts, thus, $\frac{2}{5}$, and called two-fifths; and so on. In this method, the unit may be divided into any number of equal parts, and any number of such parts may be denoted.

With this elementary view of fractions, it may be perceived, that when there is a remainder of 1 unit, it is to be divided into as many equal parts as there are units in the divisor, and one of these parts is to be annexed to the quotient. This is performed by merely writing the 1 as a numerator, and the divisor as the denominator, on the right of the quotient. If the remainder be 2 units, there will be 2 such parts of a unit as the divisor indicates to be annexed to the quotient, and, therefore, the numerator will be 2. If the remainder be 3 units, the numerator will be 3; and so on. Hence, whatever the remainder may be, it becomes, in the quotient, the numerator of a fraction, the divisor being the denominator.

RULE FOR DIVISION. *When the divisor does not exceed 9, draw a line under the dividend, find how many times the divisor is contained in the left hand figure, or two left hand figures of the dividend, and write the figure expressing the number of times underneath: if there be a remainder over, conceive it to be prefixed to the next figure of the dividend, and divide the next figure as before. Thus proceed through the dividend.*

When the divisor is more than 9, find how many times

it is contained in the fewest figures that will contain it, on the left of the dividend, write the figure expressing the number of times to the right of the dividend, for the first quotient figure ; multiply the divisor by this figure, and subtract the product from the figures of the dividend considered. Place the next figure of the dividend on the right of the remainder, and divide this number as before. Thus proceed through the dividend. If there be a final remainder, place it as a numerator, and the divisor as a denominator, on the right of the quotient.

PROOF. *Multiply the whole numbers of the divisor and quotient together, and to the product add the numerator of any fraction in the quotient: the sum will be equal to the dividend, if the work be right.*

Abbreviations of the above rule may frequently be adopted, as follows:

When there are ciphers on the right hand of a divisor, cut them off, and omit them in the operation; also cut off and omit the same number of figures from the right hand of the dividend. Finally, place the figures cut off from the dividend, on the right of the remainder.

When the divisor is 10, 100, 1000, &c., cut off as many figures from the right hand of the dividend as there are ciphers in the divisor; the other figures of the dividend will be the quotient, and the figures cut off will be the remainder.

When factors of the divisor are known, divide the dividend by one of these factors, and the quotient thence arising by the other: the last quotient will be the true one. To find the true remainder, multiply the last remainder by the first divisor, and to the product add the first remainder.

1. Divide 4062900311 by 9, and prove the operation.

2. How many times is 502 contained in 74260710?

3. Suppose 52076348 to be a dividend, and 8649 the divisor; what is the quotient?

4. If 26537009535 be divided into 27856 equal parts, what will be one of those parts?

5. Divide 16500269842 by 86000 ; abbreviating.

6. Divide 8065743924 by 10000 ; by abbreviation.
7. Divide 290516 by 63 ; using factors of the divisor.
8. 142375800392 + 5274 = what number?

VII.

PROPERTIES OF NUMBERS.

Before proceeding to examine the properties of numbers, a few arithmetical terms, which we shall here collect and define, should be perfectly understood. As an exercise in this article, the learner may give, upon his slate, an example of each term defined, and each property described.

A UNIT, or UNITY, is any thing considered individually, without regard to the parts of which it is composed.

An INTEGER is either a unit or an assemblage of units; and a FRACTION is any part or parts of a unit.

One number is said to MEASURE another, when it divides it without leaving any remainder.

A number which divides two or more numbers without a remainder, is called their COMMON MEASURE.

When a number can be measured by another, the former is called the MULTIPLE of the latter.

If a number can be measured by two or more numbers, it is called their COMMON MULTIPLE.

A COMPOSITE NUMBER is that which can be measured by some number greater than unity.

The ALIQUOT PARTS of a number, are the parts by which it is measured, or into which it can be divided.

An EVEN NUMBER is that which can be measured, or exactly divided by 2.

An ODD NUMBER is that which cannot be measured by 2; it differs from an even number by 1.

A PRIME NUMBER is that which can only be measured by unity, that is, by 1.

One number is PRIME TO ANOTHER, when unity is the only number by which both can be measured.

A SQUARE NUMBER is the product of two equal factors; or, the product of a number multiplied by itself.

The SQUARE ROOT is the number, which, being multiplied by itself, produces the square number.

A CUBE is the product of three equal factors; or, the product of a number twice multiplied by itself.

The CUBE ROOT is the number, which, being twice multiplied by itself, produces the cube.

Property 1.   The sum, or the difference of any two even numbers, is an even number.

Prop. 2.   The sum, or difference, of two odd numbers is even; but the sum of three odd numbers is odd.

Prop. 3.   The sum of an even number of odd numbers is even; but the sum of an odd number of odd numbers is odd.

Prop. 4.   The sum, or the difference of an even number and an odd number, is odd.

Prop. 5.   The product of an even, and an odd number, or of two even numbers, is even.

Prop. 6.   An odd number cannot be divided by an even number, without a remainder.

Prop. 7.   A square number, or a cube number, arising from an even root, is even.

Prop. 8.   The product of any two odd numbers is an odd number.

Prop. 9.   The product of any number of odd numbers is odd: hence the square, and the cube of an odd number are odd.

Prop. 10.   If an odd number measure an even number, it will also measure the half of it.

Prop. 11.   If a square number be either multiplied or divided by a square, the product or quotient is a square.

Prop. 12.   If a square number be either multiplied or divided by a number that is not a square, the product or quotient is not a square.

Prop. 13.   The difference between an integral cube and its root, is always divisible by 6.

Prop. 14.   The product arising from two different prime numbers cannot be a square.

Prop. 15.　The product of no two different numbers, prime to each other, can make a square, unless each of those numbers be a square.

Prop. 16.　Every prime number above 2, is either 1 greater or 1 less than some multiple of 4.

Prop. 17.　Every prime number above 3, is either 1 greater or 1 less than some multiple of 6.

Prop. 18.　The number of prime numbers is unlimited.　The first ten are, 1, 2, 3, 5, 7, 11, 13, 17, 19, 23. The learner may find the succeeding ten.

VIII.

PROBLEMS.

A PROBLEM is a proposition or a question requiring something to be done; either to investigate some truth or property, or to perform some operation.

The following Problems and Rules are founded in the correspondence of the four principal operations of arithmetic; viz. Addition, Subtraction, Multiplication, and Division.

PROBLEM I.　The sum of two numbers, and one of the numbers being given, to find the other. *RULE. Subtract the given number from the given sum; the remainder will be the number required.*

1. Suppose 37486 to be the sum of two numbers, one of which is 8602; what is the other?

2. 33000 news-papers are sold in London, daily: of these, 17500 are morning papers, the rest, evening: how many of the latter?

PROBLEM II.　The difference between two numbers, and the greater number being given, to find the smaller. *RULE. Subtract the difference from the greater number; the remainder will be the number required.*

3. If 1406 be the difference between two numbers, and the greater number be 4879, what is the smaller?

4. The area of North and South America is 18000000 square miles: that of North America is 11000000: what is that of South America?

PROBLEM III. The difference between two numbers, and the smaller number being given, to find the greater. *RULE. Add the smaller number and the difference together; the sum will be the number required.*

5. Suppose 86974 to be the difference between two numbers, and the smaller number to be 7064; what is the greater number?

6. The British House of Lords consists of 427 members; the number in the House of Commons is 131 greater. How many are there in the House of Commons?

PROBLEM IV. The sum and difference of two numbers being given, to find the numbers. *RULE. Subtract the difference from the sum, and divide the remainder by 2; the quotient will be the smaller number. Then add the given difference to the smaller number, and this sum will be the greater number.*

7. What are the two numbers whose sum is 1094, and whose difference is 154?

8. The United States Congress, consisting of a Senate and House of Representatives, has 288 members. The House has 192 members more than the Senate. How many in each branch?

PROBLEM V. The product of two factors, and one of the factors being given, to find the other. *RULE. Divide the product by the given factor, and the quotient will be the required factor.*

9. 1246038849 is the product of some two numbers, one of which is 269181: what is the other?

10. Suppose a session of Congress which continues 180 days, to cost 504000 dollars; what is the expense per day, to the United States?

PROBLEM VI. The dividend and quotient being given

to find the divisor. *RULE. Divide the dividend by the given quotient, and the quotient thence arising will be the number sought.*

11. Suppose 101442075 to be a dividend, and 4025 the quotient; what is the divisor?

12. 17155 pounds of beef having been equally divided among a number of soldiers, each one found that his share was 47 pounds. What was the number of soldiers?

PROBLEM VII. The divisor and quotient being given, to find the dividend. *RULE. Multiply the divisor and quotient together; the product will be the required dividend.*

13. If 800027 be a divisor, and 97563 the quotient, what number is the dividend?

14. A quantity of beef was divided equally among 2742 soldiers, and each soldier received for his share 152 pounds. What quantity was divided?

PROBLEM VIII. The product of three factors, and two of those factors being given, to find the third factor. *RULE. Find the product of the two given factors, and by this number divide the given product; the quotient will be the factor required.*

15. Suppose the product of three factors to be 1344, one of these factors being 12, and another 8; what is the third factor?

16. How many days will 9720 pounds of hay last 12 horses; allowing each horse to eat 45 pounds a day?

PROBLEM IX. Two numbers being given, to find their greatest common measure; that is, the greatest number which will divide them both without a remainder. *RULE. Divide the greater number by the smaller, and this divisor by the remainder, and thus continue dividing the last divisor by the last remainder, till nothing remains. The divisor last used will be the number required.*

When the greatest common measure of more than two numbers is required, *first, find the greatest common measure of any two of the numbers, then find the greatest common measure of the number found and another of the given numbers, and thus proceed, till all the given numbers are brought in.*

17. What is the greatest common measure of 918, 1998, and 522 ?

```
918)1998(2              54)522(9
    1836                   486
    ─────                  ─────
    162)918(5              36)54(1
        810                   36
        ─────                 ───
        108)162(1             18)36(2
            108                  36
            ─────
            54)108(2
               108        •        Ans. 18.
```

The truth of the rule in this problem will be discovered by retracing the first of the above operations, as follows. Since 54 [the last divisor] measures 108, it also measures $108 + 54$, or 162. Again, since 54 measures 108 and 162, it also measures $5 \times 162 + 108$, or 918. In the same manner it will be found to measure $2 \times 918 + 162$, or 1998. Therefore, 54 measures both 918 and 1998. It is also the *greatest* common measure; for, suppose there be a greater— then, since the greater measures 918 and 1998, it also measures the remainder, 162; and since it measures 162 and 918, it also measures the remainder 108; in the same manner it will be found to measure the remainder, 54; that is, the greater measures the less, which is absurd.

18. What is the greatest common measure of the numbers, 323 and 425 ?

19. What is the greatest common measure of 2310 and 4626 ?

20. What is the greatest common measure of 1092, 1428, 1197 and 805 ?

21. Suppose a hall to be 154 feet long, and 55 wide; what is the length of the longest pole, that will exactly measure both the length and width of the hall ?

22. A owns 720 rods of land, B owns 336 rods, and C 1736 rods. They agree to divide their land into equal house lots, fixing on the greatest number of rods for a lot,

that will allow each owner to lay out all his land. How
many rods must there be in a lot?

PROBLEM X. Two or more numbers being given, to
find their least common multiple; that is, the least number that will contain each of the given numbers a whole
number of times. *RULE. Divide two or more of the
given numbers by any prime number that will measure
them, repeat the operation upon the quotients and undivided numbers, and thus continue, till they become prime
to each other. Multiply the several divisors, the last
quotients, and undivided numbers together; the product
will be the least common measure.*

If, among the numbers to be divided, any number is a
measure of another, the measuring number may be re-
jected; that is, dropped from the operation.

It is obvious, that one number is the multiple of another,
when the former contains all the factors of the latter.
The factors of 6 are 3 and 2, and the factors of 9 are 3
and 3. Now 54 contains all these factors, $(3 \times 2 \times 3 \times 3 = 54)$, and 54 is a common multiple of 6 and 9, but it
is not their *least* common multiple—it is 3 times as great
as the *least*, owing to the existence of the factor, 3, in
both 6 and 9. Hence we observe, that a *common factor*
of two or more numbers must enter but once into the
multiplication, to give the least common multiple. The
above rule effects the necessary exclusion.

23. What is the least common multiple of 12, 25, 30,
and 45.

3)	12	25	30	45
5)	4	25	10	15
	4	5	2	3

$3 \times 5 \times 4 \times 5 \times 3 = 900$

We find, after dividing
twice, that 4 and 2 ap-
pear; and, by dropping
the 2 because it measures
the 4, we avoid another
division. *Ans.* 900.

24. What is the least common multiple of 6, 10, 16,
and 20?
25. What is the least common multiple of 25, 35, 60,
and 72?

26. What is the least common multiple of 105, 140, and 245 ?

27. What is the least common multiple of 18, 82, 94, 788, and 356 ?

28. Allowing 63 gallons to fill a hogshead, 42 a tierce, and 32 a barrel, what is the smallest quantity of molasses, that can be first shipped in some number of full hogsheads, then discharged and reshipped in some number of full tierces, and again discharged and reshipped in some number of full barrels ?

29. A certain flour dealer, who purchased his flour from a mill on the opposite side of a river, owned four boats, one of which would carry 8 barrels of flour, another 9, another 15, and another 16. What is the smallest number of barrels he could purchase, that would make some number of full freights for either of the boats ?

IX.

COMPOUND NUMBERS.

COMPOUND NUMBERS are those which are employed to express quantities that consist of several denominations; each denomination being denoted separately. Under this head are classed, all the subdivisions of measures; of length, surface, solidity, weights, money, time, &c.

The following tables of denominations of compound numbers, show how many units of each lower denomination are equal to a unit of the next higher, and, exhibit each lower denomination as a fraction of the next higher.

MONEY, WEIGHTS, AND MEASURES.

ENGLISH MONEY.

The denominations of English Money are, the pound, £, the shilling, *s.*, the penny, *d.*, and the farthing, *qr.*

4 farthings	$= 1$ d.	1 qr. $= \frac{1}{4}$ of 1 d.
12 pence	$= 1$ s.	1 d. $= \frac{1}{12}$ of 1 s.
20 shillings	$= 1$ £.	1 s. $= \frac{1}{20}$ of 1 £.

3

TROY WEIGHT.

The denominations of Troy Weight are, the pound, *lb.*, the ounce, *oz.*, the pennyweight, *dwt.*, and the grain, *gr.*

24 grains .... = 1 dwt.	1 gr. ... = $\frac{1}{24}$ of 1 dwt.
20 pennyweights ... = 1 oz.	1 dwt. .. = $\frac{1}{20}$ of 1 oz.
12 ounces .... = 1 lb.	1 oz. ... = $\frac{1}{12}$ of 1 lb.

AVOIRDUPOIS WEIGHT.

The denominations of Avoirdupois Weight are, the ton, *T.*, the hundred-weight, *cwt.*, the quarter, *qr.* the pound, *lb.*, the ounce, *oz.*, and the dram, *dr.*

16 drams .... = 1 oz.	1 dr. ... = $\frac{1}{16}$ of 1 oz.
16 ounces .... = 1 lb.	1 oz. ... = $\frac{1}{16}$ of 1 lb.
28 pounds .... = 1 qr.	1 lb. ... = $\frac{1}{28}$ of 1 qr.
4 quarters .... = 1 cwt.	1 qr. ... = $\frac{1}{4}$ of 1 cwt.
20 hundred-weight . = 1 T.	1 cwt. .. = $\frac{1}{20}$ of 1 T.

APOTHECARIES' WEIGHT.

The denominations of Apothecaries' Weight are, the pound, ℔, the ounce, ℥, the dram, ʒ, the scruple, ℈, and the grain, *gr.*

20 grains .... = 1 ℈.	1 gr. .... = $\frac{1}{20}$ of 1 ℈.
3 scruples .... = 1 ʒ.	1 ℈ ..... = $\frac{1}{3}$ of 1 ʒ.
8 drams .... = 1 ℥.	1 ʒ ..... = $\frac{1}{8}$ of 1 ℥.
12 ounces .... = 1 ℔.	1 ℥ ..... = $\frac{1}{12}$ of 1 ℔.

CLOTH MEASURE.

The denominations of Cloth Measure are, the French ell, *Fr. e.*, the English ell, *E. e.*, the Flemish ell, *Fl. e.*, the yard, *yd.*, the quarter, *qr.*, and the nail, *na.*

4 nails .... = 1 qr.	1 na. .. = $\frac{1}{4}$ of 1 qr.
4 quarters ...... = 1 yd.	1 qr. .. = $\frac{1}{4}$ of 1 yd.
3 quarters ...... = 1 Fl. e.	1 qr. .. = $\frac{1}{3}$ of 1 Fl. e.
5 quarters ...... = 1 E. e.	1 qr. .. = $\frac{1}{5}$ of 1 E. e.
5 quarters ...... = 1 Fr. e.	1 qr. .. = $\frac{1}{5}$ of 1 Fr. e.

DRY MEASURE.

The denominations of Dry Measure are, the bushel, *bu.*, the peck, *pk.*, the gallon, *gal.*, the quart, *qt.*, and the pint, *pt.*

2 pints .... = 1 qt.	1 pt. .... = $\frac{1}{2}$ of 1 qt.
4 quarts .... = 1 gal.	1 qt. .... = $\frac{1}{4}$ of 1 gal.
8 quarts .... = 1 pk.	1 qt. .... = $\frac{1}{8}$ of 1 pk.
4 pecks .... = 1 bu.	1 pk. .... = $\frac{1}{4}$ of 1 bu.

WINE MEASURE,

The denominations of Wine Measure are, the ton, *T.*, the pipe, *p.*, the puncheon, *pun.*, the hogshead, *hhd.*, the tierce, *tier.*, the barrel, *bl.*, the gallon, *gal.*, the quart, *qt.* the pint, *pt.* and the gill, *gi.*

4 gills . . . . = 1 pt.	1 gi. . . = $\frac{1}{4}$ of 1 pt.
2 pints . . . . . . = 1 qt.	1 pt. . . = $\frac{1}{2}$ of 1 qt.
4 quarts . . . . . = 1 gal.	1 qt. . . = $\frac{1}{4}$ of 1 gal.
31$\frac{1}{2}$ gallons . . . . = 1 bl.	1 gal. . . = $\frac{2}{63}$ of 1 bl.
42 gallons . . . . . = 1 tier.	1 gal. . . = $\frac{1}{42}$ of 1 tier.
63 gallons . . . . . = 1 hhd.	1 gal. . . = $\frac{1}{63}$ of 1 hhd.
84 gallons . . . . . = 1 pun.	1 gal. . . = $\frac{1}{84}$ of 1 pun.
126 gallons . . . . . = 1 p.	1 gal. . . = $\frac{1}{126}$ of 1 p.
2 pipes . . . . . = 1 T.	1 p. . . = $\frac{1}{2}$ of 1 T.

BEER MEASURE.

The denominations of Beer Measure are, the butt, *bt.*, the hogshead, *hhd.*, the barrel, *bl.*, the kilderkin, *kil.*, the firkin, *fir.*, the gallon, *gal.*, the quart, *qt.*, and the pint, *pt.*

2 pints . . . . = 1 qt.	1 pt. . . . = $\frac{1}{2}$ of 1 qt.
4 quarts . . . . . . = 1 gal.	1 qt. . . . = $\frac{1}{4}$ of 1 gal.
9 gallons . . . . . . = 1 fir.	1 gal. . . . = $\frac{1}{9}$ of 1 fir.
2 firkins . . . . . . = 1 kil.	1 fir. . . . = $\frac{1}{2}$ of 1 kil.
2 kilderkins . . . . = 1 bl.	1 kil. . . . = $\frac{1}{2}$ of 1 bl.
3 kilderkins . . . . = 1 hhd.	1 kil. . . . = $\frac{1}{3}$ of 1 hhd.
2 hogsheads . . . . . = 1 bt.	1 hhd. . . = $\frac{1}{2}$ of 1 bt.

NOTE. In the United States, the Dry gallon contains 268$\frac{4}{5}$ cubic inches, the Wine gallon 231 cubic inches, and the Beer gallon 182 cubic inches. By an Act of the British government, however, the distinction between the Dry, Wine, and Beer gallon was abolished in Great Britain, in 1826, and an *Imperial Gallon* was established, as well for liquids as for dry substances. The Imperial gallon must contain "10 pounds, Avoirdupois weight, of distilled water, weighed in air, at the temperature of 62° of Fahrenheit's thermometer, the barometer standing at 30 inches." This quantity of water will be found to measure 277$\frac{274}{1000}$ cubic inches. The same Act establishes the *pound Troy* at 5760 grains, and the *pound Avoirdupois* at 7000 grains.

LONG MEASURE.

The denominations of Long Measure are, the mile, *m.*, the furlong, *fur.*, the rod or pole, *r.*, the yard, *yd.*, the foot, *ft.*, and the inch, *in.*

12 inches ...... = 1 ft.	1 in. ... = $\frac{1}{12}$ of 1 ft.
3 feet .... = 1 yd.	1 ft. ... = $\frac{1}{3}$ of 1 yd.
5½ yards ...... = 1 r.	1 yd. ... = $\frac{2}{11}$ of 1 r.
40 rods .... = 1 fur.	1 r. ... = $\frac{1}{40}$ of 1 fur.
8 furlongs ..... = 1 m.	1 fur. .. = $\frac{1}{8}$ of 1 m.

SQUARE MEASURE.

The superficial contents of any figure having four sides and four equal angles, is found in squares, by multiplying together the length and breadth of the figure.

The denominations of Square Measure are, the mile, *m.*, the acre, *A.*, the rood, *R.*, the rod, *r.*, the yard, *yd.*, the foot, *ft.*, and the inch, *in.*

144 inches ...... = 1 ft.	1 in. .. = $\frac{1}{144}$ of 1 ft.
9 feet .... = 1 yd.	1 ft. ... = $\frac{1}{9}$ of 1 yd.
30¼ yards ...... = 1 r.	1 yd. .. = $\frac{1}{121}$ of 1 r.
40 rods .... = 1 R.	1 r. ... = $\frac{1}{40}$ of 1 R.
4 roods ...... = 1 A.	1 R. .. = $\frac{1}{4}$ of 1 A.
640 acres ...... = 1 m.	1 A. .. = $\frac{1}{640}$ of 1 m.

CUBIC MEASURE.

The cubical contents of any thing which has 6 sides—its opposite sides being equal—is found in cubes, by multiplying together, the length, breadth and depth.

The denominations of Cubic Measure are, the yard, *yd.*, the foot, *ft.*, and the inch, *in.*

1728 inches ..... = 1 ft.	1 in. .. = $\frac{1}{1728}$ of 1 ft.
27 feet ...... = 1 yd.	1 ft. .. = $\frac{1}{27}$ of 1 yd.

40 feet of round timber, or 50 feet of hewn timber make a *ton.* 16 cubic feet make a *foot of wood*, and 8 feet of wood make a *cord.*

TIME.

The denominations of Time are, the year, *Y.*, the day, *d.*, the hour, *h.*, the minute, *m.*, and the second, *s.*

60 seconds ...... = 1 m.	1 s. .... = $\frac{1}{60}$ of 1 m.
60 minutes ..... = 1 h.	1 m. ... = $\frac{1}{60}$ of 1 h.
24 hours .... = 1 d.	1 h. .... = $\frac{1}{24}$ of 1 d.
365 days .... = 1 Y.	1 d. .... = $\frac{1}{365}$ of 1 Y.

The earth revolves round the sun once in 365 days, 5 hours, 48 minutes, 48 seconds: this period is therefore a *Solar* year. - In order to keep pace with the solar year, in our reckoning, we make every fourth year to contain 366 days, and call it Leap year. Still greater accuracy requires, however, that the Leap day be dispensed with 3 times, in every 400 years. Whenever the number which denotes the year can be measured by 4, the year is Leap year— the centurial years excepted.

The year is also divided into 12 months—See Almanac.

THE CIRCLE.

The divisions of the circle, *C.*, are, the sign, *S.*, the degree, (°), the minute, ('), the second, (").

This table is applied to the Zodiac; and by it are computed, planetary motions, latitude, longitude, &c.

60 seconds ...... = 1'	1" ..... = $\frac{1}{60}$ of 1'
60 minutes ...... = 1°	1' ..... = $\frac{1}{60}$ of 1°
30 degrees ...... = 1 S.	1° ..... = $\frac{1}{30}$ of 1 S.
12 signs .... = 1 C.	1 S. .... = $\frac{1}{12}$ of 1 C.

GEOGRAPHICAL MEASURE.

The circumference of the globe—like every other circle—is divided in 360 equal parts, called *degrees*. Each degree is divided into 60 equal parts called *miles*, or *minutes*. Three miles are called a *league*.

On the equator, $69\frac{1}{6}$ statute miles are equal to 60 geographical miles, or 1 degree, nearly: and, on the meridian, at a mean, $69\frac{1}{12}$ statute miles are equal to a degree.

REDUCTION OF COMPOUND NUMBERS.

REDUCTION is the operation of changing any quantity from its number in one denomination, to its number in another denomination.

RULE FOR REDUCTION. *When a greater denomination is to be reduced to a smaller, multiply the greater denomination, by that number which is required of the smaller, to make a unit of the greater; adding to the product, so many of the smaller denomination as are expressed in the given quantity. Perform a like operation on this product, and on each succeeding product.*

3*

When a smaller denomination is to be reduced to a greater, divide the smaller denomination by that number which is required of the smaller, to make a unit of the next greater: the quotient will be of the greater denomination, and the remainder will be of the same denomination with the dividend. Perform a like operation on this quotient, and on each succeeding quotient.

1. Reduce £351 13s. 0d. 1qr. to its value in farthings.
2. How many pounds, &c. are there in 6169 pence ?
3. In 59lb. 13dwt. 5gr. Troy, how many grains ?
4. Change 20571005 drams to its value in tons, &c.
5. In 231℔ 3℥ 0℈ 0℈ 5gr. how many grains ?
6. How many English ells are there in 352 nails ?
7. Reduce 7 bushels and 6 quarts to pints.
8. How many hhds. are there in 9576 pints of wine ?
9. How many pints in 1bl. 1fir. 1pt. of beer ?
10. How many miles, &c. are there in 26431 rods ?
11. In 3 square miles, how many square rods ?
12. In 1259712 cubic inches, how many cubic yards ?
13. Reduce 1 solar year, 7d. and 10h. to seconds.

ADDITION OF COMPOUND NUMBERS.

The operation of adding compound numbers, differs from that of adding simple numbers, only, with respect to the irregular system of units, which determines the principles of carrying from one denomination to another.

RULE. *Write the numbers so that each denomination shall stand in a separate column. Add the numbers of the lowest denomination together, and divide their sum by that number which is required of this denomination to make a unit of the next higher: write the remainder under the column added, and carry the quotient to the next column. Thus proceed through the denomination.*

14. What is the sum of £9 8s. 4d., £250 8s. 5d. 3qr., £9 7s. 4d., £20 16s. 4d., and 3s. 6d. 2qr.?
15. Add together 10oz. 14dwt. 16gr., 5lb. 9oz. 6dwt. 22gr., 4lb. 1oz. 18dwt. 9gr., and 11dwt., Troy
16. Add together 15T. 19cwt. 3qr. 2lb. 7oz., 25T. 13cwt. 2qr. 20lb. 15oz., and 3qr. 26lb.

17. How much is 18 yd. 3 qr. 3 na., 15 yd. 2 qr. 3 na., 25 yd. 1 qr. 2 na., and 57 yd. 3 qr. 2 na. of cloth?

18. Add together 25 bu. 3 pk. 7 qt., 100 bu. 2 pk. 4 qt., 215 bu. 2 pk. 2 qt. 1 pt., and 57 bu. 3 pk. of corn.

19. Add together 4 p. 125 gal. 3 qt., 75 gal. 2 qt. 1 pt., 35 p. 92 gal., and 39 gal. 3 qt. 1 pt. of wine.

20. How many acres are 13 A. 3 R. 38 r., 87 A. 2 R. 33 r., 28 A. 2 R., 41 A. 2 R. 28 r., and 36 r.?

21. How much hewn timber is 9 T. 19 ft. 1725 in., 150 T. 39 ft. 1695 in., and 500 T. 31 ft. 915 in.?

SUBTRACTION OF COMPOUND NUMBERS.

RULE. *Write the several denominations of the smaller quantity under the same denominations of the greater quantity: then, begin with the lowest denomination, and perform subtraction on each denomination separately. Whenever a number expressing a denomination in the upper line is smaller than the number under it, increase the upper-number by as many as make a unit of the next higher denomination, and consider the number of the next higher denomination in the upper line, to be 1 less than it stands.*

22. Subtract 1 ℔. 10 oz. 16 dwt. from 3 ℔., Troy.

23. From 6 T. 3 cwt. take 7 cwt. 2 qr. 15 lb., Avoir.

24. From 2 ℔ 7 ℥ take 7 ℥ 6 ʒ 2 Ɵ 5 gr., Apoth. wt.

25. Subtract 3 qr. 3 na. from 5 yd. 2 qr. 1 na. of cloth.

26. Subtract 8 bu. 1 pk. 6 qt. 1 pt. from 50 bu. of corn.

27. From 3 hhd. 25 gal. take 41 gal. 2 qt. of wine.

28. From 6 bl. 1 kil. take 1 fir. 6 gal. 3 qt. of beer.

29. Subtract 3 yd. 10 in. from 5 yd. 2 ft. 2 in., Long mea.

30. Subtract 57 A. 2 R. 31 r. from 1 m., Square mea.

31. Subtract 2 Y. 90 d. 4 h. 55 m. from 4 Y., Time.

MULTIPLICATION OF COMPOUND NUMBERS.

RULE. *Begin with the lowest denomination, and multiply each denomination separately; divide each product by the number which is required of its own denomination*

*to make a unit of the next higher; write the remainder
under the denomination multiplied, and carry the quotient
to the product of the next higher denomination.*

32. Multiply £215 19 s. 6 d. by 72 or its factors.
33. Multiply 2 lb. 5 oz. 7 dwt. 10 gr., Troy, by 56.
34. What is 16 times 18 cwt. 3 qt. 15 lb., 14 oz.?
35. What is 81 times 36 bu. 3 pk. 6 qt. 1 pt., Dry mea.
36. Multiply 4 p. 105 gal. 3 qt. of wine by 60.
37. Multiply 2 m. 7 fur. 35 r., Long mea., by 63.
38. Multiply 4 m. 320 A. 1 R. 9 r., Square mea., by 15.
39. Multiply 2 Y. 250 d. 14 h. 30 m., Time, by 96.

DIVISION OF COMPOUND NUMBERS.

RULE. *Divide each denomination separately, beginning with the highest. Whenever a remainder occurs,
reduce it to the next lower denomination, add it to the
number expressed in the lower denomination, and divide
it therewith.*

40. Divide £251 15 s. 7 d. 2 qr. into 46 equal parts.
41. Divide 15 lb. 3 oz. 7 dwt. 5 gr., Troy, by 13.
42. Divide 12 T. 27 lb. 15 oz., Avoirdupois, by 5.
43. Divide 136 E. e. 3 qr. 3 na. of cloth by 31.
44. Divide 1621 bu. 2 pk. of corn into 50 equal parts.
45. Divide 1 pipe of wine equally among 9 owners.
46. Divide a Leap year into 100 equal parts.

FEDERAL MONEY.

The denominations of Federal Money are, the eagle,
the dollar, the dime, the cent, and the mill. 10 mills
make 1 cent, 10 cents 1 dime, 10 dimes 1 dollar, and 10
dollars 1 eagle.. Dollars, $; and Cents, *cts.* are the only
denominations commonly mentioned in business— eagles
being counted as tens of dollars, dimes being counted as
tens of cents, and mills not being denoted.

100 cents . . . . = $1 ‖ 1 cent . . . = $\frac{1}{100}$ of $1

The cents in any number of dollars are expressed by
the same figures which express the dollars, with two

ciphers annexed; $15 = 1500 cents. The dollars in any number of cents are distinguished by cutting off two figures from the right for cents; 325 cts. = $3.25.

Operations on numbers expressing Federal money, are performed as on simple numbers; care must however be taken, in addition and subtraction, to place dollars under dollars, and cents under cents; these denominations being separated by a point.

47. What is the sum of $34.21, $7064.04, 36 cts., $10004.85, $96, $900.10, $14, $1.99, and $76529?

48. Subtract $4926 from $12262.37.

49. Subtract $297.18 from $100000.

50. Suppose $295.48 to be a multiplicand, and 25 the multiplier; what is the product?

In multiplication, only one of the factors can be Federal money, and the product will be of the same denomination as this factor. If, therefore, there be cents in either factor, two figures must be pointed off for cents, from the right of the product.

51. What is the product of 96 cts. multiplied by 43?

52. What is the value of 1304 pounds of coffee at 9 cents per pound?

53. How many times $7 are there in $29.46?

In division, when both the dividend and divisor are Federal money, they must both be of the same denomination. If therefore, one of the numbers contain cents, and the other dollars only, the latter number must have two ciphers annexed to it.

54. How many barrels of flour, at $4.36 per barrel, can be purchased for $4370?

55. Divide $4279.50 into 746 equal parts.

56. If 407 pounds of Hyson tea cost $395, what is the cost of 1 pound?

57. How many times are 95 cts. contained in $56?

58. A merchant sold 1248 yards of cloth, at such price as to gain 1 cent on every nail. How much did he gain?

59. What is the gain on a hogshead of molasses, sold at an advance of 3 cents per gallon?

60. A jeweller sold a silver pitcher 3 lb. 8 oz. 16 dwt., at 7 cents a pennyweight. What did it amount to?

61. What is the freight of 60480 pounds of cotton from Charleston to Liverpool, at $4 per ton?

MISCELLANEOUS EXAMPLES.

62. How many bottles, holding 1 pint and 2 gills each, are required for bottling 4 barrels of cider?

63. How much will 46 bushels of oats cost, at 4 pence 2 farthings for every two quarts?

64. A brewer sold 96 hogsheads of beer for £388 16 s. What was the price of 1 pint at the same rate?

65. A certain tippler spent 12 cents a day for ardent spirit, during 39 successive weeks, and then died, the victim of his folly. What did the spirit all cost?

66. Bought five loads of wood; the first containing 1 cord 32 cubic feet, the second 1 cord 64 cubic feet, the third 112 cubic feet, the fourth 1 cord 28 cubic feet, and the fifth 1 cord 20 cubic feet. How many cords were there in the whole?

67. Bought goods to the amount of £25 13 s. 10 d. 2 qr.; and afterwards sold goods to the same man, amounting to £30 10 s. 4 d. 2 qr. What is the balance of money in my favor?

68. A farmer sold five lots of land, at $9 an acre; the first lot containing 30 A. 2 R. 2 r., the second 41 A. 3 R. 8 r., the third 14 A. 1 R. 10 r., the fourth 25 A. 36 r., and the fifth 54 A. 6 r. What did the whole amount to?

69. How many cubic inches in a brick 8 inches long, 4 inches wide, and 2 inches thick?

70. How many cubic inches in the cube of 2 inches? . . . . . in the cube of 3 inches? . . . . . in the cube of 4 inches? . . . . . in the cube of 5 inches?

71. If the cube of 4 inches be taken from the cube of 1 foot, how many cubic inches will remain?

72. If the cube of 4 inches be taken from the cube of 2 feet, how many cubic inches will remain?

73. A young man, on commencing business, was worth £643 10 s.; the first year he cleared £54 11 s. 7 d. 2 qr.; the second year, £87 0 s. 10 d. 1 qr.; but the third year he lost £196 7 s. 11 d. 3 qr. How much was he then worth?

74. A gentleman had a hogshead of wine in his cellar, from which there leaked out 17 gal. 3 qt. 1 pt.   How much then remained ?

75. A man started on a journey of 20 miles 6 fur. 29 r., and stopped to rest at a house, 4 m. 4 fur. 20 r. from the place of starting.   How far had he still to go ?

76. In a pile of wood, 96 feet long, 5 feet high, and 4 feet wide, how many cords ?

77. How much would 13 hogsheads of sugar cost, at 8 cents per pound; allowing each hogshead to contain, 8 cwt. 3 qr. 24 lb. ?

78. A cent weighs 8 pennyweights 16 grains.   What is the weight of 100 cents ?

79. How many yards of cloth are there in 19 pieces; each piece containing 27 yd. 3 qr. 2 na. ?

80. If a man sell 2 bl. 1 kil. 1 fir. 6 gal. 2 qt. 1 pt. of beer in one week, how many barrels would he sell in 26 weeks ?

81. If 1 pint and 3 gills of wine will fill a bottle, how much will fill a gross, or 12 dozen bottles ?

82. A father left an estate worth £ 5719 17 s., to be divided equally among 11 children.   How much was each one's share ?

83. Sixteen men own 24 tierces of molasses, in equal shares.   What is one man's share ?

84. A company of 23 men bought 1850 acres 10 rods of wild land, and divided it equally among them.   How much land had each man ?

85. What must be the length of a lot of land, that is 5 rods wide, in order that the lot shall contain 1 acre ?

Observe in the above question, that 1 acre contains 160 square rods; and, that this number of square rods is the product of the two factors that denote the width and length of the lot.   See PROBLEM V, page 21.

86. What must be the depth of a house lot, that measures 72 feet on the front, to contain 9432 square feet ?

87. What must be the length of a stick of hewn timber, that is 10 inches wide and 1 ft. 3 in. deep, in order that the stick shall contain 1 ton ?

Observe in this question, that the number of cubic

inches in a ton, is the product of the three factors which denote, in inches, the width and depth and length of the stick. See PROBLEM VIII, page 22.

88. What must be the length of a pile of wood that is 4 feet wide and 3 feet high, in order that the pile shall contain 1 cord, that is, 128 cubic feet?

89. Suppose a pile of wood to be 11 feet long and 3 feet wide; how high must it be, to contain 2 cords 4 feet of wood and 10 cubic feet?

X.

FRACTIONS.

A FRACTION signifies one or more of the equal parts into which a unit, or some quantity considered as an integer, or whole, is divided.

A fraction is expressed by two numbers or *terms*, written one above the other, thus, $\frac{3}{4}$. The lower term —called the *denominator*—denotes the number of equal parts into which the integer is divided; and the upper term—called the *numerator*—indicates what number of those equal parts the fraction expresses.

We may not only consider a fraction as a certain number of parts of a unit, but, may also view it as a part of a certain number of units. Thus, $\frac{2}{3}$ may either be considered as 2-thirds of 1, or, 1-third of 2; for 1-third of 2 is the same quantity as 2-thirds of 1. Hence, if the numerator of a fraction be viewed as an integer, and divided into as many equal parts as the denominator indicates, the fraction may be regarded as expressing *one* of these parts. Thus, if 4 be divided into 5 equal parts, the fraction $\frac{4}{5}$ expresses one of these parts.

Fractions generally have their origin from the division of a number by another which does not measure it; the excess of the dividend, above what can be measured by the divisor, being the numerator, and the divisor being the denominator, as shown in ART. VI.

If the numerator of a fraction be made equal to the

denominator, the fraction becomes equal to unity; thus $\frac{4}{4}=1$. If the numerator be greater than the denominator, the fraction is equal to as many units as the denominator is contained times in the numerator; for example $\frac{12}{4}=3$. Hence, a fraction may be viewed as an *unexecuted division*; the divisor being written under the dividend. It follows, also, that since any number divided by 1 gives the same number in the quotient, any number may be expressed as a fraction by making 1 its denominator. For example, 17 may be expressed thus, $\frac{17}{1}$.

The following propositions concerning fractions, should be distinctly noticed.

PROPOSITION I. *As many times as the numerator is made greater, so many times the fraction is made greater; and, as many times as the numerator is made smaller, so many times the fraction is made smaller. Hence, a fraction is multiplied by multiplying the numerator, and divided by dividing the numerator.*

PROPOSITION II. *As many times as the denominator is made greater, so many times the fraction is made smaller; and as many times as the denominator is made smaller, so many times the fraction is made greater. Hence, a fraction is divided by multiplying the denominator, and multiplied by dividing the denominator.*

PROPOSITION III. *When the numerator and denominator are both multiplied, or both divided by the same number, the quantity expressed by the fraction is not thereby changed.*

A PROPER FRACTION is a fraction whose numerator is less than its denominator; as $\frac{2}{13}$.

An IMPROPER FRACTION is a fraction whose numerator equals, or exceeds its denominator; as $\frac{9}{9}$, $\frac{15}{4}$.

A number consisting of an integer with a fraction annexed, as $14\frac{7}{8}$, is called a MIXED NUMBER.

A COMPOUND FRACTION is a fraction of a fraction; as $\frac{3}{4}$ of $\frac{1}{2}$. $\frac{1}{3}$ of $\frac{7}{12}$ of $\frac{2}{3}$.

A COMPLEX FRACTION is that which has a fraction either in its numerator, or in its denominator, or in both of them; thus, $\frac{5\frac{1}{2}}{6}$, $\frac{8}{9\frac{1}{4}}$, $\frac{4\frac{1}{2}}{7\frac{1}{2}}$.

4

REDUCTION OF FRACTIONS.

REDUCTION OF FRACTIONS consists in changing them from one form to another, without altering their value.

CASE I. To reduce a fraction to its lowest terms; that is, to change the denominator and numerator to the smallest numbers that will express the same quantity.

RULE. *Divide both terms of the fraction by their greatest common measure, and the two quotients will be the lowest terms of the fraction.* See PROB. IX, page 22.

When the greatest common measure is readily perceived, the fraction may be reduced mentally. For instance, the greatest common measure of the terms of the fraction $\frac{8}{12}$, is 4, and the only notation necessary in the reduction, is, $\frac{8}{12} = \frac{2}{3}$.

Dividing the terms of a fraction by a common measure which is *not the least*, will reduce it in some degree, and when thus reduced, it may be reduced still lower by another division, and so on, till no number will measure both the terms. For example, to reduce $\frac{12}{42}$, divide by 2, and the result is $\frac{6}{21}$; again, divide by 3, and the result is $\frac{2}{7}$. Here the fraction is known to be in its lowest terms, because the terms are prime to each other.

1. Reduce $\frac{384}{1152}$ to its lowest terms, by repeatedly dividing the terms by any common measure.

2. Reduce $\frac{918}{1998}$ to its lowest terms, by dividing the terms by their greatest common measure.

3. Reduce each of the following fractions to its lowest terms. $\frac{32}{120}$. $\frac{270}{306}$. $\frac{384}{1152}$. $\frac{156}{336}$. $\frac{720}{1736}$. $\frac{3108}{3552}$.

CASE II. To reduce a whole number to an improper fraction.

RULE. *Multiply the whole number by the proposed denominator, and the product will be the numerator.*

When the quantity to be reduced is a *mixed number*, the numerator of the fraction in the mixed number must be added to the product of the whole number, and their sum will be the numerator of the improper fraction.

4. Reduce 16 to a fraction whose denominator is 9.

$$\begin{array}{r} 16 \\ 9 \\ \hline 144 \end{array} \quad Ans. \ \tfrac{144}{9}$$

In 1 unit there are 9-ninths; therefore, there are 9 times as many ninths as there are units in any number.

5. Reduce 75 to a fraction whose denominator is 13.
6. Reduce 3 to a fraction whose denominator is 342.
7. How many fifteenths are there in 74 ?
8. How many eighths of a dollar in $647 ?
9. Reduce $36\tfrac{4}{7}$ to an improper fraction.

$$\begin{array}{r} 36\tfrac{4}{7} \\ 7 \\ \hline 256 \end{array} \quad Ans. \ \tfrac{256}{7}$$

In this example, we add the 3-sevenths to the sevenths produced by the multiplication of 36 by 7, and thus obtain $\tfrac{256}{7}$.

10. Reduce $25\tfrac{13}{24}$ to an improper fraction.
11. Reduce $615\tfrac{3}{234}$ to an improper fraction.
12. How many sixteenths of a dollar in $541\tfrac{9}{16}$?

CASE III. To reduce an improper fraction to a whole number, or a mixed number.

RULE. *Divide the numerator by the denominator, and the quotient will be the whole, or mixed number.*

13. Reduce $\tfrac{362}{8}$ to a whole, or mixed number.

$$\begin{array}{r} 8)362 \\ \hline 45\tfrac{2}{8} = 45\tfrac{1}{4} \end{array}$$

Since $\tfrac{8}{8}$ are equal to 1 unit, there are as many units in $\tfrac{362}{8}$ as there are times 8 in 362.

14. Reduce $\tfrac{4891}{27}$ to a whole, or mixed number.
15. How many units are there in $\tfrac{45315}{15}$?
16. How many dollars in $\tfrac{262}{7}$ of a dollar ?

CASE IV. To reduce a compound fraction to a simple, or single fraction.

RULE. *Multiply all the numerators together for a new numerator, and all the denominators for a new denominator: then reduce the new fraction to its lowest terms.*

When any numerator is equal to any denominator, the operation may be abbreviated by rejecting both.

If part of the compound fraction be an integer, or a mixed number, it must first be reduced to an improper fraction.

17. Reduce $\frac{1}{2}$ of $\frac{2}{3}$ of $\frac{3}{4}$ of 6 to a simple fraction.

$\frac{1}{2}\times\frac{2}{3}\times\frac{3}{4}\times\frac{6}{1}=\frac{12}{32}=\frac{3}{8}$ | Here the common term, 3, is omitted in the multiplication.

18. Reduce $\frac{5}{2}$ of $\frac{2}{11}$ to a simple fraction.
19. Reduce $\frac{1}{4}$ of $\frac{6}{15}$ of $\frac{12}{20}$ to a simple fraction.
20. Reduce $\frac{11}{24}$ of $\frac{3}{137}$ of $2\frac{5}{11}$ to a simple fraction.
21. Reduce $\frac{1}{2}$ of $\frac{2}{3}$ of $\frac{3}{4}$ of 5 to a simple fraction.

CASE V. To reduce a fraction from one denomination to another.

RULE. *Multiply the proposed denominator by the numerator of the given fraction, and divide the product by the denominator of the given fraction; the quotient will be the numerator of the proposed denominator.*

22. Reduce $\frac{5}{6}$ to a fraction whose denominator shall be 14: or, in other words change 5-sixths to fourteenths.

$\begin{array}{r}14\\5\\\hline 6)70\\\hline 11\frac{4}{6}\end{array}$ *Ans.* $\frac{11\frac{4}{6}}{14}$ | $\frac{1}{6}$ is equal to $\frac{1}{6}$ of $\frac{14}{14}$, and $\frac{5}{6}$ is 5 times as much: we therefore find 5 times 14-fourteenths and take $\frac{1}{6}$ of this product for the required fourteenths.

23. How many fifths are there in $\frac{7}{8}$?
24. $\frac{4}{13}$ is equal to how many twenty-fourths ?
25. Reduce $\frac{3}{8}$ to a fraction whose denominator is 4.
26. How many twelfths of 1 shilling in $\frac{4}{7}$ of 1 s.?

CASE VI. To reduce the lower denominations of a compound number to the fraction of a higher denomination.

RULE. *Reduce the given quantity to the lowest denomination mentioned, and this number will be the numerator: then reduce a unit of the higher denomination to the same denomination with the numerator, and this number will be the denominator.*

27. Reduce 7 oz. 18 dwt. 13 gr. to the fraction of a pound.

We find, that 7 oz. 18 dwt. 13 gr. when reduced to grains, gives 3805 for the numerator; and 1 pound when reduced to grains, gives 5760 for the denominator. Therefore, $\frac{3805}{5760}=\frac{261}{1152}$ is the fraction required.

28. Reduce 4 s. 9 d. 3 qr. to the fraction of £1.

29. Reduce $3\frac{1}{2}$ inches to the fraction of a yard.
30. What fraction of a hogshead is 9 gal. $2\frac{4}{5}$ pt.?
31. Reduce 5 cwt. 8 lb. 4 oz. to the fraction of a ton.

CASE VII. To reduce the fraction of a higher denomination to its value in whole numbers of lower denomination.

RULE. *Multiply the numerator by that number of the next lower denomination which is required to make a unit of the higher, and divide the product by the denominator; the quotient will be a whole number of the lower denomination, and the remainder will be the numerator of a fraction. Proceed with this fraction as before, and so on.*

It will be readily perceived, that the fraction of a higher denomination is reduced to the fraction of a lower, by rhultiplying the numerator by the number of units of the lower, required to make a unit of the higher. Thus, $\frac{1}{5}$ of a bushel is 4 times as many fifths of a peck; that is, $\frac{4}{5}$ of a peck. Again, $\frac{4}{5}$ of a peck is 8 times 12-fifths, that is, $\frac{96}{5}$ of a quart; and again, $\frac{96}{5}$ of a quart is 2 times 96-fifths, that is, $\frac{192}{5}$ of a pint. If the denominator be multiplied, instead of the numerator, the effect is the reverse, and the fraction is reduced to a higher denomination. Thus, $\frac{3}{5}$ of a pint, (the 5 being multiplied by 2,) becomes $\frac{3}{10}$ of a quart; $\frac{3}{10}$ of a quart, (the 10 being multiplied by 8,) becomes $\frac{3}{80}$ of a peck; and $\frac{3}{80}$ of a peck, (the 80 being multiplied by 4,) becomes $\frac{3}{320}$ of a bushel.

32. Reduce $\frac{11}{12}$ of a gallon to its value in quarts, &c.

$$11$$
$$4$$
$$12)\overline{44}$$
$$\overline{3}\quad 8$$
$$2$$
$$12)\overline{16}$$
$$\overline{1}\quad 4$$
$$4$$
$$12)\overline{16}$$
$$\overline{1\frac{4}{12}}=1\frac{1}{3}$$

We find by multiplication, that $\frac{11}{12}$ of a gallon is $\frac{44}{12}$ of a quart; and, by division, that $\frac{44}{12}$ of a quart is 3 qt. and $\frac{8}{12}$ of a quart. We then find, that $\frac{8}{12}$ of a qt. is $\frac{16}{12}$ of a pint; and, that $\frac{16}{12}$ of a pt. is 1 pt. and $\frac{4}{12}$ of a pt. And thus, by finding the units of one denomination at a time, we finally obtain the whole answer, which, denoted as a compound number, is 3 qt. 1 pt. $1\frac{1}{3}$ gi.

33. Reduce $\frac{4}{5}$ of £ 1 to its value in shillings &c.

34. Reduce $\frac{1}{2}\frac{1}{4}$ of a yard, to its value in feet, &c.

35. In $\frac{7}{16}$ of 1 cwt. how many quarters, pounds, &c.?

36. Reduce $\frac{12}{20}$ of a bushel to pecks, quarts, and pints.

CASE VIII. To reduce fractions to a common denominator; that is, to change two or more fractions which have different denominators, to equivalent fractions, that shall have the same denominator.

RULE 1st. *Multiply each numerator into all the denominators except its own, for a new numerator. Then multiply all the denominators together for a new denominator, and place it under each new numerator.*

RULE 2nd. *Find the least common multiple of the given denominators for the common denominator; then divide the common denominator by each given denominator and multiply the quotient by its given numerator; the several products will be the several new numerators.* (See PROBLEM X, page 23.)

The 1st. of the above rules is convenient when the terms of the fractions are small numbers, but the 2nd. is otherwise to be preferred, as it always gives a denominator which is the least possible. Other methods of finding a common denominator will occur to the student, after further practice.

If any of the fractions to be reduced to a common denominator be *compound*, they must first be simplified.

37. Reduce $\frac{5}{8}$, $\frac{11}{12}$, $\frac{9}{14}$ and $\frac{13}{15}$ to a common denominator.

In this example, the least common denominator is found to be 840. Then the several numerators of the common denominator are found as follows.

$840 \div 8 = 105$, and $105 \times 5 = 525$. *Ans.* $\frac{5}{8} = \frac{525}{840}$

$840 \div 12 = 70$, and $70 \times 11 = 770$. $\frac{11}{12} = \frac{770}{840}$

$840 \div 14 = 60$, and $60 \times 9 = 540$. $\frac{9}{14} = \frac{540}{840}$

$840 \div 15 = 56$, and $56 \times 13 = 728$. $\frac{13}{15} = \frac{728}{840}$

38. Reduce $\frac{3}{17}$, $\frac{2}{15}$ and $\frac{4}{25}$ to a common denominator.

39. Reduce $\frac{5}{8}$, $\frac{19}{17}$, $\frac{1}{2}$ and $\frac{5}{6}$ to a common denominator.

40. Reduce $\frac{27}{61}$ and $\frac{12}{8}$ to a common denominator.

41. Reduce $\frac{1}{3}$ and $\frac{1}{4}$ of $\frac{5}{4}$ to a common denominator.

CASE IX. To reduce a complex fraction to a simple fraction.

RULE. *If the numerator or denominator, or both, be whole or mixed numbers, reduce them to improper fractions: multiply the denominator of the lower fraction into the numerator of the upper, for a new numerator; and multiply the denominator of the upper fraction into the numerator of the lower, for a new denominator.*

42. Reduce $\dfrac{9}{3\frac{1}{7}}$ to a simple fraction.

The operation. $\dfrac{9}{3\frac{1}{7}} = \dfrac{\frac{9}{1}}{\frac{22}{7}} = \dfrac{9 \times 7}{1 \times 22} = \dfrac{63}{22}$ *Ans.* $\dfrac{63}{22}$

43. Simplify each of the following complex fractions.

$$\dfrac{4}{\frac{1}{7}}. \quad \dfrac{\frac{4}{7}}{5}. \quad \dfrac{5\frac{1}{7}}{8}. \quad \dfrac{3\frac{1}{4}}{5\frac{1}{3}}. \quad \dfrac{21}{4\frac{1}{3}}. \quad \dfrac{6\frac{2}{3}}{7\frac{3}{11}}. \quad \dfrac{2\frac{2}{3}}{1\frac{1}{2}}.$$

ADDITION OF FRACTIONS.

Fractions are added by merely adding their numerators, but they must be of the same integers; we cannot *immediately* add together $\frac{5}{8}$ of a yard and $\frac{3}{8}$ of an inch, for the same reasons that we cannot immediately add together 5 yards and 3 inches. They must, also, be of the same denomination; we cannot immediately add together *fourths* and *fifths*.

RULE. *Reduce compound fractions, (if there be any), to simple fractions, and reduce all to a common denominator; then add together the numerators, and place their sum over the common denominator. If the result be an improper fraction, reduce it to a whole or mixed number.*

44. Add together, $3\frac{7}{9}$, $\frac{5}{8}$, $8\frac{5}{8}$ and $\frac{3}{4}$.

	$\overline{360}$	
$3\frac{7}{9}$	280	
$\frac{5}{8}$	225	
$8\frac{5}{8}$	216	
$\frac{3}{4}$	270	
$13\frac{271}{360}$	$\frac{991}{360} = 2\frac{271}{360}$	

By operations not here denoted, we find the common denominator to be 360; and also find the several new numerators. The sum of the fractions is $\frac{991}{360} = 2\frac{271}{360}$, which, added to the whole numbers, gives the total sum, $13\frac{271}{360}$

45. Add together, $9\frac{2}{7}$, $12\frac{5}{14}$, $\frac{4}{15}$, $\frac{3}{8}$ and $21\frac{3}{4}$.

46. What is the sum of $\frac{9}{10}+\frac{3}{8}+\frac{4}{7}+\frac{2}{7}+\frac{8}{21}$?

47. What is the sum of $19\frac{5}{12}+\frac{7}{8}$ of $\frac{5}{6}+2\frac{6}{15}+\frac{4}{18}$?

48. What is the sum of $\frac{1}{2}$ of $\frac{12}{20}+37+6\frac{7}{9}+\frac{6}{11}$?

49. Find the sum of $\frac{7}{9}$ of a shilling and $\frac{2}{5}$ of a penny?

In this example, first reduce the $\frac{7}{9}$ of a shilling to pence, and the fraction of a penny.

50. Find the sum of $\frac{2}{13}$ of a gallon and $\frac{7}{8}$ of a gill.

51. What is the sum of $5\frac{5}{6}$ days and $52\frac{5}{15}$ minutes?

52. What is the sum of $\frac{4}{7}$ of a cwt., $8\frac{2}{3}$ lb. and $3\frac{9}{40}$ oz.?

SUBTRACTION OF FRACTIONS.

As in addition of fractions we find the sum of their numerators, so in subtraction of fractions we find the *difference* of their numerators.

RULE. *If either quantity be a compound fraction, reduce it to a simple fraction, and if the two fractions have different denominators, reduce them to a common denominator. Subtract the numerator of the subtrahend from the numerator of the minuend, and place the remainder over the common denominator.*

When the minuend is a mixed number, and the fraction in the subtrahend is greater than that in the minuend, subtract the numerator of the subtrahend from the denominator, and to the difference add the numerator of the minuend; and consider the integer of the minuend to be 1 less than it stands.

It is not always obvious, which of two fractions expresses the greater quantity. In such case, the fractions are denoted with a character between them, thus, $\frac{15}{22} \smile \frac{11}{17}$; and the greater is discovered by reducing them to a common denominator.

53. What is the difference between $24\frac{7}{9}$ and $26\frac{3}{8}$?

$$26\frac{3}{8}$$
$$24\frac{7}{9}$$
$$\overline{1\frac{43}{72}}$$

$$72$$
$$\overline{27}$$
$$56$$
$$\overline{\frac{43}{72}}$$

Here the fraction in the subtrahend is the greater, and we are obliged to convert a unit into *seventy-seconds* to obtain a quantity from which to subtract $\frac{56}{72}$.

54. What is the difference between $\frac{9}{14}$ from $\frac{16}{19}$?

55. Perform subtraction on $\frac{15}{22}$ ∽ $\frac{11}{17}$.
56. What will remain if $51\frac{7}{12}$ be taken from $84\frac{2}{5}$?
57. Subtract $\frac{7}{8}$ of $\frac{1}{4}$ from $36\frac{9}{10}$.
58. What is the difference between $4\frac{1}{15}$ and $10\frac{1}{16}$?
59. What will remain if $\frac{2}{3}$ of $\frac{7}{8}$ be taken from a unit?
60. What is the difference between $\frac{32}{271}$ and $\frac{15}{74}$?
61. $4\frac{2}{3} - \frac{1}{2}$ of $\frac{3}{4}$ of $\frac{2}{3}$ is equal to what quantity?

MULTIPLICATION OF FRACTIONS.

The following rules for multiplication of fractions, are based on the Propositions I, and II, stated in page 37.

CASE I.   To multiply a fraction by a whole number.
RULE.   *Either multiply the numerator, or divide the denominator by the whole number.*

CASE II.   To multiply a whole number by a fraction.
RULE.   *Multiply the whole number by the numerator, and divide the product by the denominator.*

CASE III.   To multiply a fraction by a fraction.
RULE.   *Multiply numerator by numerator, and denominator by denominator, for a new fraction.*

When both factors are mixed numbers, it is generally more convenient to reduce them to improper fractions and then proceed according to the rule under Case III.

The effect of multiplying any quantity by a proper fraction is, to give in the product, such a part of the quantity multiplied as the fraction indicates. Thus the product must be less than the multiplicand. This effect of the operation will appear consistent with the principle of multiplication, when it is considered, that multiplying any number by 1, gives only the same number in the product; and, therefore, multiplying by *less* than 1, must give a product *less* than the number multiplied.

62. Multiply $\frac{25}{28}$ by 9. $\frac{25}{28} \times 9 = \frac{25 \times 9}{28} = \frac{225}{28} = 8\frac{1}{28}$
63. Multiply 49 by $\frac{5}{7}$.   (See rule under Case II.)
64. Multiply $\frac{7}{12}$ by $\frac{1}{4}$.   (See rule under Case III.)
65. Multiply $6\frac{9}{10}$ by $3\frac{4}{13}$.   (Remark under Rule III.)
66. What is the product of $\frac{11}{28}$ by 15?

67. What is the product of 9241 by $\frac{3}{10}$?

68. What is the product of $\frac{5}{12}$ by $\frac{22}{33}$?

69. What is the product of $8\frac{3}{10}$ by $12\frac{9}{16}$?

70. Which is the most, $\frac{7}{8} \times 65$, or, $65 \times \frac{7}{8}$?

71. What is the product of $294\frac{1}{2}$ by 25?

In this example, it will be most convenient to find the product of the whole numbers without regard to the fraction first; then find the product of the fraction in a separate operation, and, finally, add the two products together.

72. What is the product of 361 by $34\frac{1}{2}$?

73. How many square inches of paper in a sheet that is $14\frac{7}{8}$ inches long, and $11\frac{2}{3}$ inches wide?

DIVISION OF FRACTIONS.

The rules for division of fractions, like those for multiplication, are based on Propositions I, and II.

CASE I.   To divide a fraction by a whole number.

RULE. *Either divide the numerator, or multiply the denominator, by the whole number.*

CASE II.   To divide a whole number by a fraction.

RULE. *Multiply the whole number by the denominator, and divide the product by the numerator.*

CASE III.   To divide a fraction by a fraction.

RULE. *Invert the divisor, and then proceed as in multiplying a fraction by a fraction.*

Observe, that the operation of this last rule is, to multiply the denominator of the dividend by the numerator of the divisor for a new denominator, and the numerator of the dividend by the denominator of the divisor for a new numerator.

'Compound fractions are to be reduced to simple ones; and mixed numbers to improper fractions, before the adoption of either of the above rules.

74. Divide $\frac{5}{12}$ by 8. $\frac{5}{12} \div 8 = \frac{5}{12 \times 8} = \frac{5}{96}$. *Ans.*

75. Divide 14 by $\frac{7}{15}$.   (See rule under Case II.)

76. Divide $\frac{5}{24}$ by $\frac{2}{3}$.   (See rule under Case III.)

77. Divide the compound fraction $\frac{1}{2}$ of $\frac{9}{10}$ by 6.

78. Divide 325 by the mixed number $5\frac{3}{4}$.

79. What is the quotient of $\frac{2\frac{1}{2}}{4}$ divided by 13?

80. What is the quotient of 57 divided by $\frac{7}{12}$?

81. What is the quotient of $\frac{7}{18}$ divided by $\frac{3\frac{1}{2}}{15}$?

82. Divide $\frac{7}{8}$ of $\frac{9}{10}$ by $\frac{2}{3}$ of $\frac{1}{4}$ of $\frac{5}{6}$.

83. What is the quotient of $91\frac{2}{3}$ divided by 15?

84. What is the quotient of $206\frac{3}{5}$ divided by $9\frac{4}{17}$?

85. How many times is $\frac{2\frac{1}{4}}{4}$ contained in 319?

86. How many times is $19\frac{1}{5}$ contained in $99\frac{1}{2}$?

87. How many times $\frac{7}{8}$ of an inch in $\frac{9}{10}$ of a yard?
First, reduce the $\frac{9}{10}$ of a yd. to the fraction of an inch.

88. How many times $\frac{2}{5}$ of a gill in 3 barrels?

89. Suppose a wheel to be $11\frac{5}{14}$ feet in circumference; how many times will it roll round in going $39\frac{3}{7}$ rods?

MISCELLANEOUS EXAMPLES.

In the following examples, all fractions which appear in the answers, must be reduced to their value in whole numbers of lower denominations, whenever there is opportunity for such reduction.

90. What distance will a car run in $9\frac{3}{4}$ hours, allowing its velocity to be $23\frac{5}{8}$ miles an hour?

91. Suppose a car wheel to be 8 feet 7 inches in circumference, how many times will it turn round in running $46\frac{1}{2}$ miles?

92. If $3\frac{7}{8}$ cwt. of sugar be taken from a hogshead containing 14 cwt. 1 qt. $6\frac{1}{2}$ lb., how much will remain in the hogshead?

93. What is the sum of $16\frac{2}{3}$ cwt., 7 cwt. 3 qr. $8\frac{1}{4}$ lb., 2 T. $19\frac{4}{5}$ cwt., 2 cwt. $1\frac{1}{2}$ qr., and $\frac{7}{8}$ of a ton?

94. A farmer owning $132\frac{4}{5}$ acres of land, sold 46 A. 3 R. 12 r. How much land had he remaining?

95. What is the value of $36\frac{3}{7}$ acres of land, at $\$47\frac{5}{8}$ per acre?

96. What is the value of $15\frac{1}{2}$ barrels of flour, at $\$4.62\frac{1}{2}$ per barrel?

97. What is the value of a load of wood, containing 6 feet, [$\frac{6}{8}$ of a cord,] at $\$5.25$ per cord? Or, what is $\frac{6}{8}$ of $\$5.25$? Or, $\$5.25 \times \frac{6}{8} = ?$

98. How much land is there in a square lot, measuring $354\frac{1}{2}$ rods on every side? (See page 28.)

99. What quantity of land in a lot, which is $65\frac{1}{2}$ rods long and $47\frac{3}{4}$ rods wide?

100. What quantity of wood is there in a pile, $14\frac{4}{12}$ feet long, $3\frac{9}{12}$ feet wide, and $6\frac{3}{12}$ feet high?

101. Suppose a lot of land to be $6\frac{1}{4}$ rods wide, how long must it be, to contain 1 acre? (See PROB. V, page 21. Consider that 1 acre contains 160 rods.)

102. What quantity of loaf sugar must be sold at $19\frac{3}{10}$ cents per pound, that the price shall amount to $\$524$?

103. What cubical quantity of earth must be removed, in digging a pit, $13\frac{1}{2}$ feet deep, $12\frac{1}{4}$ feet long, and $9\frac{2}{3}$ feet wide?

104. What quantity of hewn timber is there in a stick that is $12\frac{1}{2}$ feet long, $2\frac{1}{4}$ feet deep, and $1\frac{3}{4}$ foot wide?

105. Suppose a stick of timber to be $1\frac{1}{12}$ foot deep, and 8 inches wide; what must be the length of the stick, in order that its quantity shall be 1 ton of hewn timber? (See PROB. VIII, page 22. Consider a ton as the product of three factors.)

106. Suppose wood to be piled on a base 18 feet long and $7\frac{3}{4}$ feet wide, what must be the height of the pile, to contain $9\frac{1}{4}$ cords?

107. What quantity of molasses in 4 casks, containing severally, $55\frac{1}{2}$ gal., $31\frac{3}{8}$ gal., $27\frac{2}{15}$ gal., and $58\frac{3}{17}$ gal.?

108. What is the cost of $486\frac{3}{4}$ bushels of corn, at $62\frac{1}{2}$ cents per bushel?

109. Suppose $6\frac{2}{5}$ gallons to have leaked from a hogshead of wine, what is the value of the remainder of the wine, at $87\frac{1}{2}$ cents per gallon?

110. How many bottles, each holding $1\frac{1}{2}$ pint, are required for bottling 3 barrels of cider?

111. Suppose $4\frac{1}{3}$ gallons of cider to have evaporated from a barrel; what number of bottles, each holding 1 pt. $3\frac{1}{3}$ gl., will be required to bottle the remainder?

112. What is the value of $142\frac{1}{2}$ tons of coal, at $7\frac{3}{5}$ dollars per ton?

113. What is the value of $\frac{7}{8}$ of a bushel of wheat, at the rate of $\frac{2}{5}$ of a dollar per bushel? $[\frac{2}{5}\times\frac{7}{8}=?]$

114. If 1 hogshead [63 gal.] of molasses cost $26¾, what is the cost of 1 gallon?

115. What is the cost of 7 hhd. 6½ gal. molasses, at 11 4/10 cents per gallon?

116. What is the cost of 25 yards 3½ quarters of ribbon, at 19¼ cents per yard?

117. If 5¼ cords of wood cost $26½, what is the cost of 1 cord?

118. What is the value of 16⅔ tons of hay, at 11⅛ dollars per ton?

119. What is the value of 1 lb. 6 oz. 12 dwt. of silver, at 20¼ cents per pennyweight?

120. If 16¾ yards of broad-cloth cost $86.24, what is the cost of 1 yard?

121. At 5s. 3¼d. per yard, what is the cost of 78¾ yards of cambric, in pounds, shillings, and pence?

122. If 492¾ yards of cloth cost £68 4s. 10d., what is the cost of 1 yard?

123. If 18¾ yards of cotton cost 12s. 9d., what is the cost of 1 yard?

124. What is the value of 5768½ lb. of coffee at 10¾ pence per pound?

125. At what price per pound must I sell 432½ pounds of coffee, in order to receive £27 3s. for the whole?

126. If £448 7/10 be equally divided among 76 men, what will each man receive?

127. If ⁴⁄₇ of a yard of cloth cost $3, what is the price of 1 yard? Or, $3 ÷ ⁴⁄₇ = ?

128. If 7 13/16 barrels of apples cost $21½, what is the cost of 1 barrel of the apples?

129. If 4½ gallons of molasses cost $2⅖, what is the cost of 1 quart?

130. If 1⅛ hogshead of wine cost $250⅓, what is the cost of 1 quart?

131. Bought 5 yards of silk, at $2¼ per yard; 15¼ yards of ribbon, at 12½ cents per yard; 17 pairs of gloves, at 68¼ cents per pair; and 16¼ yards of lace, at $3¼ per yard. What is the whole cost?

132. Bought 6¼ pounds of tea, at 87½ cents per pound; 15½ pounds of sugar, at 11½ cents per pound; 13¼ pounds

of coffee at $12\frac{1}{2}$ cents per pound; and $16\frac{1}{2}$ gallons of molasses, at $\frac{1}{3}$ of a dollar per gallon. What is the whole cost?

133. Bought $9\frac{1}{2}$ barrels of cider, at \$$2\frac{1}{3}$ per barrel; 8 barrels of apples, at \$ $1\frac{2}{3}$ per barrel; 16 boxes of raisins, at \$$2.62\frac{1}{2}$ per box; $23\frac{1}{4}$ pounds of almonds, at $14\frac{3}{4}$ cents per pound. What is the whole cost?

134. Bought $358\frac{1}{4}$ bushels of wheat, at $\frac{7}{8}$ of a dollar per bushel; 420 bushels of rye, at $96\frac{1}{3}$ cents per bushel; $146\frac{1}{4}$ bushels of corn, at $\frac{5}{8}$ of a dollar per bushel; and $651\frac{1}{8}$ bushels of oats, at $23\frac{2}{3}$ cents per bushel. What is the whole cost?

135. A purchased of B, $75\frac{5}{6}$ tons of iron at \$$9.61\frac{1}{2}$ per ton. What quantity of coffee, at $12\frac{1}{4}$ cents per pound, must A sell B, to cancel the price of the iron?

136. C purchased of D, 1397 hogsheads of molasses, at $15\frac{3}{8}$ cents per gallon; and D, at the same time, purchased of C, $896\frac{1}{4}$ tons of iron, at \$ $9\frac{1}{3}$ per ton. How much was the balance— and to whom was it due?

137. What is the sum of $\frac{14}{15}$, $\frac{7}{12}$, $\frac{6}{7}$, $\frac{21}{24}$, $\frac{16}{19}$, $\frac{18}{41}$, $\frac{9}{11}$, $\frac{4}{5}$, $\frac{3}{5}$, $\frac{7}{9}$, $\frac{46}{47}$, $\frac{37}{4}$, and $\frac{7}{10}$ of $\frac{3}{8}$?

138. Suppose $\frac{9}{10}$ of $\frac{6}{15}$ of $\frac{7}{14}$ to be a minuend, and $\frac{7}{8}$ of $\frac{2}{3}$ of $\frac{1}{5}$ of $\frac{4}{9}$ a subtrahend; what is the remainder?

139. What is the product of $\frac{1}{2}$ of $\frac{2}{3}$ of $\frac{3}{4}$ of 100, multiplied by $\frac{4}{5}$ of $\frac{5}{6}$ of $\frac{7}{8}$ of $\frac{8}{9}$ of 75?

140. What is the quotient of $\frac{4}{5}$ of $\frac{7}{8}$ of $\frac{19}{20}$, divided by $\frac{1}{5}$ of $\frac{14}{16}$ of $\frac{6}{9}$ of $\frac{12}{23}$ of $\frac{5}{8}$?

141. Suppose the sum of two fractions to be $\frac{5}{8}$, and one of the fractions to be $\frac{9}{20}$; what is the other? (See Problem I, page 20.)

142. Suppose the greater of two fractions to be $\frac{14}{19}$, and their difference to be $\frac{13}{20}$; what is the smaller fraction? (See Prob. II, page 20.)

143. Suppose the smaller of two fractions to be $\frac{24}{41}$, and their difference to be $\frac{3}{25}$; what is the greater fraction? (See Prob. III, page 21.)

144. What are the two fractions, whose sum is $\frac{14}{16}$, and whose difference is $\frac{6}{27}$? (See Prob. IV, page 21.)

145. If $\frac{93}{198}$ be the product of two factors, one of which is $\frac{6}{11}$, what is the other? (See Prob. V, page 21.)

146. Suppose $\frac{7}{9}$ to be a dividend, and $\frac{4}{33}$ a quotient; what is the divisor? (See Prob. vi, page 21.)

147. What must be that dividend, whose divisor is $\frac{2\frac{1}{4}}{4}$ and whose quotient is $\frac{3}{7}$? (See Prob. vii, page 22.)

148. Suppose the product of three factors to be $\frac{12}{20}$, one of those factors being $\frac{1}{6}$, and another $\frac{2}{13}$; what is the the third factor? (See Prob. viii, page 22.)

149. A merchant owning $\frac{9}{16}$ of a ship, sold $\frac{3}{7}$ of what he owned. What part of the whole ship did he sell?

150. A merchant owning $\frac{12}{30}$ of a ship, sold $\frac{7}{8}$ of what he owned. What part of the ship did he still own?

151. If I buy $\frac{5}{18}$ of $\frac{7}{9}$ of a ship, and sell $\frac{2}{3}$ of what I bought, what part of the ship shall I have left?

The kind of fractions, which have been treated in this article, are called *Vulgar fractions*, or *Common fractions*, in distinction from another kind, called *Decimal* fractions, or simply *Decimals*.

XI.

DECIMAL FRACTIONS.

A DECIMAL FRACTION is a fraction whose denominator is 10, or 100, or 1000, &c. The denominator of a decimal fraction is never written: the numerator is written with a point prefixed to it, and the denominator is understood to be a unit, with as many ciphers annexed as the numerator has places of figures. Thus, .5 is $\frac{5}{10}$, .26 is $\frac{26}{100}$, .907 is $\frac{907}{1000}$.

When a whole number and decimal fraction are written together, the decimal point is placed between them. Thus, 68.2 is $68\frac{2}{10}$, 4.87 is $4\frac{87}{100}$.

In the notation of whole numbers, any figure, wherever it may stand, expresses a quantity $\frac{1}{10}$ as great as it would express if it were written one place further to the left: and so it is in the notation of decimal fractions—the same system is continued below the place of units. The first place to the right of units is the place of *tenths*; the second, of *hundredths;* the third, of *thousandths;* the fourth, of *ten-thousandths;* and so on.

Ciphers placed on the right hand of decimal figures, do not alter the value of the decimal; because, the figures still remain unchanged in their distance from the unit's place. For instance, .5, .50, and .500 are all of equal value,— they are each equal to $\frac{1}{2}$. But every cipher that is placed on the left of a decimal, renders its value ten times smaller, by removing the figures one place further from the unit's place. Thus, if we prefix one cipher to .5, it becomes .05 $[\frac{5}{100}]$; if we prefix two ciphers, it becomes .005 $[\frac{5}{1000}]$; and so on.

To read decimal fractions—*Enumerate and read the figures as they would be read if they were whole numbers, and conclude by pronouncing the name of the lowest denomination.*

1. Read the several numbers in the following columns.

.99	.2008	4.008	24.09
.064	.00006	6.37002	630.1174
.0003	.03795	.99999	6.972479
.5237	.130009	5.0001	28.797

2. Write in decimals the following mixed numbers.

$18\frac{7}{10}$	$2\frac{25}{1000}$	$33\frac{17}{100}$	$8\frac{11}{10000}$
$24\frac{9}{100}$	$326\frac{13}{100}$	$8\frac{201}{10000}$	$47\frac{1}{100000}$
$38\frac{6}{1000}$	$7\frac{21}{1000}$	$97\frac{42}{1000}$	$6\frac{251}{10000}$
$65\frac{8}{10000}$	$19\frac{342}{10000}$	$6\frac{1251}{10000}$	$55\frac{291}{10000000}$

ADDITION OF DECIMALS.,

3. Add the following numbers into one sum. 151.7 +70.602+4.06+807.2659.

151.7	In arranging decimals for addition,
70.602	we place tenths under tenths, hun-
4.06	dredths under hundredths, &c. We
807.2659	then begin with the lowest denomi-
1033.6279	nation, and proceed to add the col-
	umns as in whole numbers.

4. What is the sum of $256.94 + 9121.7 + 8.3065$?

5. Add together $.6517 + 19.2 + 2.8009 + 51.0007 + .00009 + 22.206 + 4.732$.

In Federal Money, the dollar is the unit; that is, dollars are whole numbers; dimes are tenths, cents are hundredths, and mills are thousandths.

6. Add together $18.25, $4.09, $2.40, $231.075, $64.207, $50.258, $10.09 and 25 cts.

7. Write the following sums of money in the form of decimals, and add them together. $1 and 1 cent, 37 cents, $25 and 7 dimes, 65 cents, $15, 9 dimes, 8 mills, 4 cents and 3 mills, $\frac{8}{10}$ of a mill, $7 and 8 cents, $\frac{6}{10}$ of a mill, $36\frac{4}{10}$ cents, 10 eagles and 25 dollars, and 7 cents.

SUBTRACTION OF DECIMALS.

8. Subtract 4.16482 from 19.375.

$$\begin{array}{r} 19.375 \\ 4.16482 \\ \hline 15.21018 \end{array}$$

After placing tenths under tenths, &c., we subtract as in whole numbers. The blank places over the 2 and 8 are viewed as ciphers.

9. Subtract 592.64 from 617.23169.

10. Subtract 48.06 from 260.3.

11. Subtract .89275 from 12690.2.

12. Subtract .281036 from 51.

13. What is the difference between 1 and .1?

14. What is the difference between 24.367 and 13?

15. What is the difference between .136 and .1295?

16. Write 8 dollars and 7 cents in decimal form, and subtract therefrom, 48 cents and 1 mill.

17. Subtract 9 dimes and 6 mills from 15 dollars.

MULTIPLICATION OF DECIMALS.

Multiplying by any fraction, is taking a certain part of the multiplicand for the product; consequently, multiplying one fraction by another, must produce a fraction smaller than either of the factors. For example, $\frac{4}{10} \times \frac{3}{10} = \frac{12}{100}$; or, decimally, $.4 \times .3 = .12$. Hence observe, that the number of decimal figures in any product, must be equal to the number of decimal figures in both the factors of that product.

RULE. *Multiply as in whole numbers; and in the product, point off as many figures for decimals, as there are decimal places in both factors. If the number of figures in the product be less than the number of decimal places in both factors, prefix ciphers to supply the deficiency.*

18. Find the product of 658 by .249.   7.06 by 3.65. .593 by 5.62.   .146 by .244.

658	7.06	5.93	.146
.249	3.65	.562	.244
5922	3530	1186	584
2632	4236	3558	584
1316	2118	2965	292
173.842	25.7690	3.33266	.035624

19. Multiply 428 by .27; that is, find .27 of 428.

20. What is the product of 3.067 by 8.2?

21. What is the product of .6247 by 23?

22. What is the product of .099 by .04?

23. What is the product of .113 by .0647?

24. What is 7.03 × .9 × 31.6 × 28.758 = ?

25. Multiply 9 dolls. 7 cts. 6 mills [9.076] by 46.

26. What cost 28 yards of cloth, at $7.515 per yd.?

27. What cost 15.9 yd. of cloth, at $9.427 per yd.?

28. What cost 275 lemons, at 9 mills apiece?

29. At 7 cents and 3 mills per yard, what is the value of 18704 yards of satin ribbon?

30. What is the value of a township containing 30519.75 acres of land, at 4 dolls. 8 cts. and 5 mills per acre?

31. What is .06 of 1532 dollars? Or, what is the product of 1532 multiplied by .06?

32. What is 03 of 476 dollars and 78 cents?

33. If an insurance office charge .015 of the value of a house for insuring it against fire, what will be the expense of insuring a house, valued at $437.25?

34. Multiply 26.000375 by .00007.

35. What is the product of 3.62981 by 10000.

The learner will perceive, that any decimal number is multiplied by 10, 100, 1000, &c., by merely removing the decimal point as many places to the right hand as there are ciphers in the multiplier. Thus, $6.25 \times 10 = 62.5$. $6.25 \times 1000 = 6250$.

DIVISION OF DECIMALS.

It has been shown, in multiplication of decimals, that there must be as many decimal places in a product as there are in both its factors; and it follows, that, in division of decimals, there must be as many decimal places in the divisor and quotient together, as there are in the dividend. Therefore, the number of decimal places in the quotient must be equal to the difference between the number of decimal places in the dividend, and the number of decimal places in the divisor.

RULE *Divide as in whole numbers; and in the quotient, point off as many figures for decimals, as the decimal places in the dividend exceed those in the divisor; that is, make the decimal places in the divisor and quotient counted together, equal to the decimal places in the dividend.*

If there be not figures enough in the quotient to point off, prefix ciphers to supply the deficiency.

When there are more decimal places in the divisor, than in the dividend, render the places equal, by annexing ciphers to the dividend, before dividing.

After dividing all the figures in the dividend, if there be a remainder, ciphers may be annexed to it, and the division continued. The ciphers thus annexed, must be counted with the decimal places of the dividend.

36. How many times is 57.2 contained in 2406.976 ?
     57.2)2406.976(42.08

37. What is the quotient of 11.7348 by 254 ?
     254)11.7348(.0462

38. What is the quotient of 4066.2 by .648 ?
     648)4066.200(6275

39. What is the quotient of 3.672 by .81?

```
.81)3.672(4 5333+
    324
    ───
    432
    405
    ───
    270
    243
    ───
    270
    243
    ───
    270
    243
    ───
     27
```

The sign of *addition*, or *more*, here shows, that the true quotient is more than the preceding figures express. We might continue the division, but we should never arrive at a complete quotient. For the purposes of business, it is seldom necessary to extend the quotient below thousandths; but, in the following exercises, those quotients that do not terminate, may be extended to millionths.

40. How many times is 4.72 contained in 637.531?

41. What is the quotient of 2.7315 by 74?

42. What is the quotient of 409.867 by .5806?

43. What is the quotient of 125 by .1045?

44. What is the quotient of 709 by 3.574?

45. What is the quotient of 7382.54 by 6.4252?

46. What is the quotient of 715 by .3075?

47. What is the quotient of 267.15975 by 13.25?

48. What is the quotient of .0851648 by 423?

49. What is the quotient of .009 by .00016?

50. If 17 boxes of oranges cost $98.29, what is the cost of a single box?

51. If $550.725 be divided equally among 15 men, what will be each man's share?

52. If 37.5 barrels of flour be divided equally among 25 men, how much will each man have?

53. If 46.75 yards of cloth cost $251.702, what is the cost of 1 yard of the cloth?

54. Divide 3712 by 42; annexing ciphers to the remainders, until eight decimal figures are obtained in the quotient.

55. What is the quotient of 9 divided by 266?

In this example it will be necessary to annex a sufficient number of decimal ciphers to the dividend, before the operation of dividing can be commenced.

56. What is the quotient 1 divided by 8 ?

57. What is the quotient of 62 divided by 97?

58. Divide 1 by 2.   3 by 4.   10 by 12.   3 by 16. 2 by 13.   6 by 26.   14 by 15.   40 by 72.   7 by 599.

Any decimal number is divided by 10, 100, 1000, &c. by merely removing the decimal point as many places to the left hand as there are ciphers in the divisor. Thus $14.8 \div 10 = 1.48$ $14.8 \div 1000 = .0148$

REDUCTION OF DECIMALS.

CASE I. To reduce a vulgar fraction to a decimal.

RULE. *Divide the numerator by the denominator, and the quotient will be the decimal.*

59. Reduce $\frac{7}{8}$ to a decimal.

| 8)7.000 | Decimal ciphers are here annexed |
| .875 *Ans.* | to the dividend as directed in the rule for division of decimals. |

60. Reduce the fractions $\frac{1}{2}$, $\frac{3}{4}$, $\frac{1}{3}$, $\frac{7}{8}$, $\frac{11}{16}$, $\frac{19}{24}$, $\frac{3}{50}$, and $\frac{9}{1125}$ to decimals.

61. Simplify $\frac{3}{5}$ of $\frac{7}{30}$, and reduce it to a decimal.

62. Reduce $\frac{1}{4}$ of $\frac{2}{3}$ of $\frac{1}{2}$ to a decimal.

63. What is the decimal expression of $247\frac{5}{16}$?

64. Reduce $\frac{2}{3}$, $\frac{2}{11}$, and $\frac{1}{27}$ to decimals.

The learner will discover, that the above fractions, $\frac{2}{3}$, $\frac{2}{11}$, and $\frac{1}{27}$ cannot be reduced to exact decimal expressions. The quotient of 2 by 3 is .6666, &c., continually. The quotient of 2 by 11 is .181818, &c.; the same *two* figures being repeated continually. The quotient of 1 by 27 is .037037, &c.; the same *three* figures being repeated continually. Decimals of this kind are treated in the next Article, under the head of *Infinite Decimals*. For most purposes, however, *three* or *four* decimal places will express any fraction with sufficient accuracy, unless the integer of the fraction is of very high value.

CASE II. To reduce a decimal to a vulgar fraction.

RULE. *Write the decimal denominator under the decimal, and erase the decimal point: view the expression as a vulgar fraction, and reduce it to its lowest terms.*

65. Reduce .4375 to a vulgar fraction.

$.4375 = \frac{4375}{10000}$; and to reduce this fraction to its lowest terms, we divide the terms by their greatest common measure, which is 265. The result is, $\frac{7}{16}$.

66. Reduce .375 to a vulgar fraction.
67. Reduce .76482 to a vulgar fraction.
68. Reduce .510505 to a vulgar fraction.
69. Reduce .1084058 to a vulgar fraction.
70. Reduce .04608128 to a vulgar fraction.

CASE III. To reduce the lower denominations of a compound number to the decimal of a higher denomination.

RULE. *Reduce the given quantity to a vulgar fraction,* (as taught in page 40), *then reduce the vulgar fraction to a decimal.*

The decimal quotients which do not terminate, may, in the examples of this case, be extended as low as the seventh place.

71. Reduce 12 s. 6 d. 3 qr. to the decimal of a £.
72. Reduce 2 qr. 14 lb. to the decimal of a cwt.
73. Reduce 1 R. 14 rods to the decimal of an acre.
74. Reduce 13 dwt. 16 gr. to the decimal of a pound, Troy weight.
75. Reduce 1 pk. 1 pt. to the decimal of a bushel.
76. Reduce 1 bl. to the decimal of a ton of wine.
77. Reduce 4 yd. 6 in. to the decimal of a mile.
78. Reduce 5 square yards to the decimal of an acre.
79. Reduce 14 cubic feet to the decimal of a cord.
80. Reduce 21 h. 50 m. 31 s. to the decimal of a year.
81. Express £19. 13 s. 9½ d. decimally; making the £ the unit, and the *s.* and *d.* a decimal.
82. Reduce 17 hhd. 9 gal. 3 qt. 1 pt. to a decimal expression; the hogshead being the unit.
83. Reduce 15 tons, 1 qr. 14 oz. to a decimal expression; the ton being the unit.
84. Reduce 4 miles, 7 fur. 9 r. 3 yd. 6 in. to a decimal expression; the mile being the unit.
85. Reduce 25 rods, 19 yd. 7 ft. 115 in., square measure, to a decimal expression; the rod being the unit.

86. What is the value of 416 gal. 3 qt. 1 pt. of wine, at $1.359 per gallon?

In this example, first reduce the quantity of wine to a decimal expression,— the gallon being the unit— and then multiply this quantity into the price of 1 gallon: the answer will be $566.533 +. The following examples in this case are to be performed in like manner.

87. What is the value of 57 yd. 2 qr. 3 na. of cloth, at $6.78 per yard?

88. What is the value of 748½ yards of ribbon, at 9 cents 8 mills [.098] per yard?

89. What is the value of 5741 yd. 3 qr. of tape, at 7 mills [.007] per yard?

90. What is the value of 4 cwt. 1 qr. 19 lb. of raisins, at $12 per hundred-weight?

91. What is the value of 32 hhd. 22 gal. of molasses, at $19.22 per hogshead?

92. What is the value of 3 pk. 7 qt. of corn, at 75 cents per bushel?

93. What is the cost of 15 E. e. 4 qr. 3 na. of linen, at $1.15 per ell?

94. What is the cost of 7 A. 2 R. 38 r. of land, at $64.50 per acre?

95. What is the cost of 28 square rods and 260 square feet of land, at $84.25 per rod?

96. What is the cost of 29 lb. 6 oz. 8 dr. of indigo, at $3.75 per pound?

97. What is the cost of 4 qr. 3 na. of thread lace, at $4.50 per French ell.?

98. What is the value of 7 lb. 10 oz. 18 dwt. of copper, at 27 cents per pound?

99. What is the value of 11 oz. 19 dwt. 23 gr. of silver. at $15.25 per pound?

CASE IV. To reduce the decimal of a higher denomination to its value in whole numbers of lower denomination.

RULE. *Multiply the decimal by that number of the next lower denomination which makes a unit of the higher, and the product will be of the lower denomination.*

*Proceed in like manner with the decimal in each suc-
ceeding product.*

100. Reduce .769 r. to its value in yards, feet, and
inches; that is, change .769 of a rod to yards, &c.

.769
5.5
‾‾‾‾‾
3845
3845
‾‾‾‾‾
4.2295
3
‾‾‾‾‾
.6885
12
‾‾‾‾‾
8.2620

There are $5\frac{1}{2}$ times, or 5.5 times
as many yards as rods in any quantity,
whether that quantity be a whole number or a decimal: therefore, we multiply the decimal of a rod by 5.5, and
the product is 4.2295 yards. We
then multiply .3295 of a yard by 3, to
find the feet; but there is not a whole
foot in this decimal, and we proceed
to find the inches. The whole result
is, 4 yd. 0 ft. 8.262 in.

101. Reduce .775 £ to its value in shillings, &c.

102. Reduce .625 s. to its value in pence and farthings.

103. Reduce .4694 lb. Troy, to oz., dwt., &c.

104. Reduce .624 cwt. to its value in qr., lb., &c.

105. Reduce .0653 mile to its value in yd., &c.

106. Reduce .3875 A. to its value in R. and rods.

107. Reduce .0098 ton to its value in lb., oz., and dr.

108. Reduce .2083 hhd. to its value in gallons.

109. Reduce .467 cwt. to its value in qr. lb. &c.

110. Reduce £741.687 to its proper expression, in
pounds, shillings, pence, and farthings.

111. Reduce 84.704 miles to its proper expression,
in miles, furlongs, rods, yards, &c.

112. Reduce 50.742 A. to its proper expression, in
the several denominations of square measure.

EXCHANGE OF CURRENCIES.

Before the adoption of the Federal currency, merchants
in this country, kept their accounts in the denominations
of English money. The value of the Pound, however,
and consequently the value of its subdivisions, was various: that is, a pound, and consequently a shilling, signified
a greater value of money in some of the states, than in
others. Accounts are now kept, in Federal money, and

its denominations are generally used in stating prices. In
some sections of the country, however, prices are frequently mentioned in shillings and pence—a custom which
is inconvenient, and which ought to be discontinued.

In New England, Virginia, Kentucky, and Tennes-
see, $\frac{1}{6}$ of a dollar is called a shilling.

In New York and North Carolina, $\frac{1}{8}$ of a dollar is
called a shilling.

In Pennsylvania, New Jersey, Delaware, and Mary-
land, $\frac{2}{15}$ of a dollar is called a shilling.

In South Carolina and Georgia, $\frac{3}{14}$ of a dollar is called a shilling.

In Canada, $\frac{1}{5}$ of a dollar is called a shilling.

113. How many cents and mills, that is, what decimal
of a dollar, in a New-England shilling? in 2 shillings?
in 3 shillings? in 4 shillings? in 5 shillings?

114. How many cents and mills in a New-York shilling? in 2 s.? in 3 s.? in 4 s.? in 5 s.? in 6 s.? in 7 s.?

115. How many cents and mills in a Pennsylvania shilling? in 2 s.? in 3 s.? in 4 s.? in 5 s.? in 6 s.?

116. How many cents and mills in a Georgia shilling?
in 2 s.? in 3 s.? in 4 s.?

117. How many cents are there in a Canada shilling?
in 2 s.? in 3 s.? in 4 s.? in 5 s.?

To change the old currencies to Federal money.

RULE. *Reduce the pounds, if there be any, to shillings. Denote the shillings as units, reduce the pence
and farthings to the decimal of a shilling, and multiply
the whole sum by that fraction of a dollar which is equal
to one shilling.*

118. Change 13 s. 6 d., of the old currency of New
England, to Federal money.

119. Change £42 19 s. 4½d. of the old currency of
New England, to Federal money.

120. Change 13 s. 6 d., of the old currency of New
York, to Federal money.

121. Change £25 17 s. 8¼d., of the old currency of
New York, to Federal money.

122. Change 18s. 11d., of the old currency of Pennsylvania, to Federal money.

123. Change £14 7s. 6½d. of the old currency of Pennsylvania, to Federal money.

124. Change 16s. 10d., of the old currency of Georgia, to Federal money.

125. Change £54 12s. 11¼d., of the old currency of Georgia, to Federal money.

126. Change 17s. 5d., of the currency of Canada, to Federal money.

127. Change £21 9s. 3¾d., of the currency of Canada, to Federal money.

128. What is the value, in Federal money, of 9 New England shillings? 9 New York shillings? 9 Pennsylvania shillings? 9 Georgia shillings? 9 Canada shillings?

MISCELLANEOUS EXAMPLES.

Any vulgar fraction, which shall appear in the following examples, must be reduced to a decimal; and the lower denominations of compound numbers must also be reduced to decimals, before they are brought into operation. No decimal need be continued lower than *six* places. Answers to be given in decimals.

129. What is the sum of 6 tons 18cwt. 1qr., 5cwt. 3qr. 2lb., 4.093825 tons, 2qr. 27lb., 8cwt. 2qr. 4lb., and 17 tons 5cwt. 0qr. 19lb.?

130. What is the difference between 2.90843 hhd. and 4hhd. 47gal. 3qt. 1pt. of wine?

131. What is the cost of 15.179 yards of broadcloth, at $6 per yard?

132. If 57 yards of cloth cost $197, what costs 1 yd.?

133. What is the cost of 28yd. 3qr. of cloth, at $7.55 per yard?

134. If 18yd. 1qr. of cloth cost $91.16, what is the cost of 1 yard?

135. What is the cost of 25cwt. 2qr. 20lb. of hops, at $4.96 per hundred weight?

136. What is the cost of 24hhd. 15gal. of molasses, at $25.36 per hogshead?

137. What is the cost of 256 yd. 3 qr. of ribbon, at 8 cents 5 mills [.085] per yard?

138. What is the cost of 24½ yards of ribbon, at .7 cents per yard?

139. What is the cost of 3 qr. 2 na. of broadcloth, at $10.35 per yard?

140. What is the cost of 1 fir. 7 gal. 3 qt. of beer, at $3.50 per firkin?

141. What is the value of 23 grains of silver, at $14 per pound, Troy?

142. What is the value of 25 square rods of land, at $75 per acre?

143. If $238.86 be divided equally among 18 men, what will each man receive?

144. If $775 be divided equally among 8 men, what will each man receive?

145. If a man travel 73.487 miles in 15 hours, what distance does he travel in 1 hour?

146. What is ⅛ of 1142.26?

The result will be the same, whether we divide 1142. 26 by the denominator 8, (which is multiplying by ⅛), or, reduce ⅛ to a decimal and multiply this decimal into 1142.26. The former method is to be preferred; and the learner is here reminded, that the product of any decimal will be such a *fractional part* of the multiplicand as the decimal indicates.

147. What is .125 of 1142.26?

148. What is ⅙ of 2.565? (Divide by 6).

149. What is .6 of 2.565? (Multiply by .6).

150. What is $\frac{1}{55}$ of 1999.2?

151. What is .56 of 1999.2?

152. What is $\frac{1}{135}$ of 387.65?

153. What is .185 of 387.65?

154. What is $\frac{6}{100}$ of 37241 dollars?

155. What is .06 of 37241 dollars?

156. Suppose I have $5872, and pay away .06 of it; how much shall I have left?

157. A owes B $430.40 to be paid in 10 months; but B relinquishes .05 of the debt for having it paid immediately. How much does B relinquish?

158. C borrowed of D, $72.85, agreeing to pay it in 16 months, increased by .08 of itself. What was the amount to be paid?

159. What will it cost to insure a house, worth $2500, against the danger of fire, for one year, the price of insurance being .025 of the value of the house?

160. Suppose I purchase a ship for $12900, and sell it at an advance equal to .019 of the cost; for how much do I sell it?

161. How many gallons of wine can be purchased for $74, at $1.37 per gallon?

162. How many pounds of raisins can be bought for $9, at 16½ cents per pound?

163. If a man travel 5.385 miles in 1 hour, in how many hours will he travel 166 miles?

164. If 18 bushels 3 pecks of wheat grow on 1 acre, how many acres will produce 396 bushels?

165. If 3 shillings will pay for 1 bushel of barley, how many bushels will 26 shillings pay for?

166. If 5 s. 8 d. will pay for 1 bushel of wheat, how many bushels will £11 pay for?

167. If 8 s. 3 d. will pay for 1 gallon of wine, how many gallons will £18 pay for?

168. What is the value, in Federal money, of £3 17 s. 8 d., of the old currency of New England?

169. If I buy 230 pelts, in Canada, at 4 s. 3 d. apiece, for what amount Federal money must I sell the whole, in the United States, in order to gain $36.15?

170. How many square feet in a floor, that is 18.63 feet long, and 14 ft. 3 in. wide?

171. How many square feet in a board, that is 16 ft. 5 in. long, and 11 inches wide?

172. How many cubic feet in a box, that is 4 ft. 6 in. long, 3 ft. 2 in. deep, and 2 ft. 9 in. wide?

173. Goliath is said to have been 6½ cubits high, each cubit being 1 foot 7.168 inches. What was his height in feet?

174. How many square feet of paper will it take to cover the walls of a room, that is 18 ft. 9 in. long, 14 ft. 6 in. wide, and 9 ft. 3 in. high?

175 Suppose a man's property to be worth $6520, and his tax to be .02 of the value of his property; how much is his tax?

176. If a man earn one dollar and one mill per day, how much will he earn in a year?

177 What is the cost of three hundred seventy-five thousandths of a cord of wood, at four dollars per cord?

178. A has nine hundred thirty-six dollars, and B has five dollars, three dimes and one mill. How much more money has A than B?

179. A trader sold 4 pieces of cloth—the first contained 86 and 3-thousandths yards; the second, 47 and 3-tenths yards; the third, 91 and 7-hundredths yards; the fourth, 22 and 9-ten-thousandths yards. What did the whole amount to, at $7 per yard?

180. A has $31.32, B has 57\frac{7}{8}$, C has 104\frac{3}{4}$, and D has 95\frac{1}{8}$; and they agree to share their money equally. What must each relinquish, or receive?

181. Suppose a car wheel to be 2 feet 9$\frac{3}{4}$ inches in circumference; how many rods will it run, in turning round 800 times?

182. If a car run 1 mile in 3 minutes and 9 seconds, in what time will it run 18 miles?

183. Suppose the sum of two certain quantities to be 1, and one of those quantities to be .8036, what is the other? (See Prob. i, page 20.)

184. Charles and Joseph together have $4.38; of which Charles's share is 17 shillings and 3 pence. What is Joseph's share?

185. Suppose .08 to be the difference between two quantities, and the greater quantity to be 80; what is the smaller? (See Prob. ii, page 20.)

186. There is a field, 5.864 acres of which is planted with corn, and the rest, with potatoes. There is 2 A. 3 R. 10 r. more of corn than potatoes. How much is planted with potatoes?

187. Suppose 7426.1 to be the difference between two quantities, and the smaller quantity to be .93; what is the greater? (See Prob. iii, page 21.)

188. Henry has $1.355 more money than William;

6*

and William has 19s. 10½d., New England currency.
How much has Henry?

189. What are the two quantities whose sum is 290.
009, and whose difference is .99? (See Prob. IV,
page 21.)

190. If a horse and chaise cost $437.25, and the
chaise cost $67.08 more than the horse, what is the cost
of each?

191. Suppose 15675.266547 to be the product of
some two factors, one of which is 27.381; what is the
other? (See Prob. V, page 21.)

192. If a board be 1 ft. 9 in. wide, how long must it
be, to contain 26.5 square feet of surface?

193. Suppose 566.916128724 to be a dividend, and
108.273 the quotient; what is the divisor? (See Prob.
VI, page 21.)

194. 4397.4 pounds of beef was equally divided among
a number of soldiers, and each soldier received 3.49
pounds. How many soldiers were there?

195. Suppose .025 to be a divisor, and .045 the quotient; what is the dividend? (See Prob. VII, page 22.)

196. Such a quantity of bread was divided equally
among 13 sailors, as allowed each sailor 1.236 pounds.
How many pounds were divided?

197. If the product of three factors be 70.4597, the
first of those factors being 3.91, and the second 3.5, what
is the third? (See Prob. VIII, page 22.)

198. What must be the depth of a pit, that is 8 ft. 5 in.
long, and 4 ft. 3 in. wide, in order that it shall contain
231 cubic feet? (Consider 231 as a product.)

199. Suppose the bottom of a wagon to be 9 feet long,
and 4 ft. 3 in. wide; how many feet high must wood be
piled in this wagon, in order that the load shall contain
1 cord? (View the cubic feet in a cord as a product.)

200. Suppose wood to be piled on a base, 15 ft. 6 in.
long, and 7 ft. 9 in. wide, what must be the height of the
pile, to contain 16 cords?

201. If a stick of timber be 1 ft. 9 in. wide, and 1.4 ft.
deep, what must be its length, in order that the stick
shall contain 1 ton?

XII.

INFINITE DECIMALS.

Learners, who are preparing for commercial business, and who do not intend to prosecute an extensive course of mathematical studies, may omit this article, and proceed immediately to Art. XIII.

INFINITE DECIMALS are those which are understood to be indefinitely continued; either by one and the same figure perpetually repeated, or, by some number of figures perpetually recurring in the same order. For example, .444444, &c. .26262626, &c. .057057057, &c. .134913491349, &c. Decimals of this kind result from division, when the divisor and dividend are prime to each other, and the divisor contains prime numbers other than those contained in 10; that is, other than 2 and 5.

An infinite decimal which is continued by the repetition of a single figure, is called a *repeating* decimal; and the repeated figure is called the *repetend*.

An infinite decimal which is continued by the repetition of more than one figure, is called a *circulating* decimal; and the repeated period of figures is called the *circulate*, or *compound repetend*.

When other decimal figures precede the repetend or circulate, the decimal is called a *mixed* infinite decimal. For example, .8476666, &c. .38171717, &c.

A single repetend is distinguished by a point over it, thus, .3, which signifies .33333, &c. A compound repetend is distinguished by a point over its first, and last figure, thus, .849, which signifies .849849849, &c.

Similar repetends—whether single or compound—are those which begin at the same place, either before or after the decimal point. For example, .13 and .72 are similar; also, .264 and .9038 are similar; also, 3.54 and 7.36 are similar.

Dissimilar repetends are those which begin at different places. For example .6127 and .405 are dissimilar.

Conterminous repetends are those which end at the same place. For example, $.7\dot49$ and $.5\dot0\dot6$.

Similar and conterminous repetends are those which begin and end at the same places. For example, $.1\dot30\dot8$ and $.4\dot01\dot2$.

Any quotient continued by annexing decimal ciphers to the dividend, is known to be infinite, whenever a remainder occurs, that has occurred before; and the repetend is known to consist of those quotient figures which succeed the first appearance, and precede the second appearance of the recurring remainder. It may also be observed, that every quotient which does not terminate, must, at some place, repeat or circulate. This truth is evident from the consideration, that the several remainders, which precede their respective quotient figures, must all be within the series of numbers, 1, 2, 3, 4, and so on, up to the number of the divisor. Therefore, it is impossible that the number of partial divisions in any operation shall equal the number indicated by the divisor, without the recurrence of some one of the remainders.

REDUCTION OF INFINITE DECIMALS.

CASE I. To reduce a repetend to a vulgar fraction.

The observations which lead to the rule are as follows. If 1, with ciphers continually annexed, be divided by 9, the quotient will be 1s continually; that is, if $\frac{1}{9}$ be reduced to a decimal, it will produce the repetend $.\dot1$: and since $.\dot1$ is the decimal equal to $\frac{1}{9}$, $.\dot2 = \frac{2}{9}$, $.\dot3 = \frac{3}{9}$, $.\dot4 = \frac{4}{9}$, and so on, up to $.\dot9 = \frac{9}{9}$ or unity. Therefore, every single repetend is equal to a vulgar fraction, whose numerator is the repeating figure, and whose denominator is 9. Again, if $\frac{1}{99}$ be reduced to a decimal, it becomes $.\dot0\dot1$; and since $.\dot0\dot1$ is the decimal equal to $\frac{1}{99}$, $.\dot0\dot2 = \frac{2}{99}$, $.\dot0\dot3 = \frac{3}{99}$, and so on, up to $.\dot9\dot9 = \frac{99}{99}$ or unity. Again, if $\frac{1}{999}$ be reduced to a decimal, it becomes $.\dot00\dot1$, and since $.\dot00\dot1$ is the decimal equal to $\frac{1}{999}$, $.\dot00\dot2 = \frac{2}{999}$,

$.00\dot{3} = \frac{3}{999}$, and so on. This correspondence exists universally; and, therefore, any circulate—not containing an integer— is equal to a vulgar fraction, whose numera-tor is the circulating figures, and whose denominator is denoted by as many 9s as there are places in the circulate.

RULE. *Make the repetend the numerator, and for the denominator take as many 9s as there are figures in the repetend.*

When there are integral figures in the repetend, a number of ciphers equal to the number of integral figures must be annexed to the numerator.

1. Reduce $.\dot{6}$ to a vulgar fraction.

2. Reduce $.\dot{0}3\dot{7}$ to a vulgar fraction; giving the frac-tion in its lowest terms.

3. Reduce $.\dot{1}2\dot{3}$ to a vulgar fraction.

4. Reduce $.\dot{1}42857$ to a vulgar fraction.

5. Reduce $.\dot{7}6923\dot{0}$ to a vulgar fraction.

6. Reduce $\dot{2}.3\dot{7}$ to a vulgar fraction.

CASE II. To reduce a mixed infinite decimal to a vulgar fraction.

Observe, that a mixed infinite decimal consists of two parts—the finite part, and the repeating part. The finite part may be reduced as shown in Art. XI, Case I; and the repeating part, as shown in the first case of this article; observing, however, to reckon the value of the fraction obtained from the repeating part ten times less for every place occupied by the finite figures. For example, the decimal $.2\dot{6}$ is divisible into the finite decimal $.2$, and the repetend $.0\dot{6}$. Now $.2 = \frac{2}{10}$, and $.\dot{6}$ would be $=\frac{6}{9}$, if the circulation began immediately after the place of units; but since it begins after the place of tenths, it is $\frac{6}{9}$ of $\frac{1}{10}$ $=\frac{6}{90}$. Then, $.2\dot{6}$ is equal to $\frac{2}{10} + \frac{6}{90} = \frac{18}{90} + \frac{6}{90} = \frac{24}{90}$.

RULE. *To as many 9s as there are figures in the repetend, annex as many ciphers as there are finite places, for a denominator. Then, multiply the same number of 9s by the finite part of the decimal, and add the repe-tend to the product, for the numerator.*

7. What is the least vulgar fraction equal to .i̇3̇?
8. Reduce .14̇8̇ to a vulgar fraction.
9. Reduce .53̇2̇ to a vulgar fraction..
10. Reduce .81̇24̇7 to a vulgar fraction.
11. Reduce .09̇2̇ to a vulgar fraction.
12: Reduce .0084̇9713̇ to a vulgar fraction.

CASE III.   To make any number of dissimilar repetends, similar and conterminous.

Observe, that a single repetend may be represented either as a compound repetend or as a mixed decimal; thus, .6̇ = .6̇6̇6̇ = .6̇6666̇. Also, a compound repetend may be represented as a mixed decimal; thus, .2̇48̇ = .24̇824̇ = .24̇82482̇4. Also, a finite decimal may be represented as a mixed infinite decimal, by annexing ciphers as repetends; thus, .39 = .39̇0̇ = 39̇00̇ = .39̇0000̇. Hence, two or more decimals, whether repetends, circulates, or mixed decimals, may be expressed with circulating figures beginning and ending together.

RULE. *Find the least common multiple of the several numbers of decimal places in the several repetends; extend the repetend which begins lowest to as many places as the multiple has units, and make all the other repetends to conform thereto.*

13. Make 6.3̇17̇, 3.4̇5̇, 52.3̇, 191.0̇3̇, .0̇57̇, 5.3̇ and 1.35̇9̇ similar and conterminous.

6.3̇17̇ =	6.31̇73173̇1
3.4̇5̇ =	3.45̇55555̇5
52.3̇ =	52.30̇00000̇0
191.0̇3̇ =	191.03̇03030̇3
.0̇57̇ =	.05̇70570̇5
5.3̇ =	5.33̇33333̇3
1.35̇9̇ =	1.35̇99999̇9

The first repetend has 3 places; the second, 2; the fourth, 2; the fifth, 3; the sixth, 1; the seventh, 1. The least common multiple of 3, 2, 2, 3, 1, 1, is 6; therefore the similar and conterminous repetends have 6 places.

14. Make 9.8̇14̇, 1.5, 87.2̇6̇, .08̇3̇ and 124.0̇9̇ similar and conterminous.

15. Make .32̇1̇, .826̇2̇, .0̇5̇, .090̇2̇ and .6̇ similar and conterminous.

16. Make .53̇1̇, .734̇8̇, .0̇7̇ .050̇3̇ and .749 similar and conterminous.

CASE IV. To find whether a given vulgar fraction is equal to a finite, or infinite decimal; and, of how many figures the repetend will consist.

If we divide unity with decimal ciphers annexed [1.0000, &c.] by any prime number, except the factors of 10, [2 and 5], the figures in the quotient will begin to repeat as soon as the remainder is 1. And since 9999, &c. is less than 10000, &c. by 1, therefore, 9999, &c. divided by any number whatever will leave 0 for a remainder, when the repeating figures are at their period. Now, whatever number of repeating figures we have, when the dividend is 1, there will be the same number, when the dividend is any other number whatever: for the product of any circulating number, by any other given number, will consist of the same number of repeating figures as before. Take, for instance, the infinite decimal .386738673867, &c. whose repeating part is 3867. Now every repetend [3867] being equally multiplied, must produce the same product: for though these products will consist of more places, yet the overplus in each, being alike, will be carried to the next, by which means each product will be equally increased, and consequently every four places will continue alike. From these observations it appears, that the dividend may be altered at pleasure, and the number of places in the repetend will still be the same: thus, $\frac{1}{11} = .0̇9̇$, and $\frac{3}{11}$ or $\frac{1}{11} \times 3 = .2̇7̇$.

RULE. *Reduce the vulgar fraction to its lowest terms, and divide the denominator by 10, 5, or 2, as often as possible. If the whole denominator vanish in dividing, the decimal will be finite, and will consist of as many figures as there are divisions performed.*

If the denominator do not vanish, then by the last quotient divide 9999, &c. till nothing remains: the num-

ber of 9s used, will show the number of places in the repetend; which will begin after so many places of figures as there were 10s, 5s, or 2s used in dividing.

17. Is the decimal equal to $\frac{12}{112}$ finite, or infinite—and if infinite, how many places has the repetend?

$$2|112$$
$$2|\;\overline{\;56\;}$$
$$2|\;\overline{\;28\;}$$
$$2|\;\overline{\;14\;}$$
$$7)999999$$
$$142857$$

Since the denominator does not vanish in dividing by 2, the decimal is infinite: and, as six 9s are used, the repetend will consist of six figures; beginning at the fifth place, because four 2s were used in dividing.

18. Examine the fraction $\frac{1}{11}$, as above directed.
19. Examine the fraction $\frac{2}{7}$, as above directed.
20. Examine the fraction $\frac{13}{404}$, as above directed.
21. Examine the fraction $\frac{1}{8544}$, as above directed.
22. Examine the fraction $\frac{31}{32}$, as above directed.

ADDITION OF INFINITE DECIMALS.

RULE. *Make the repetends similar and conterminous, and add them together. Divide this sum by as many 9s as there are places in the repetend; denote the remainder as the repetend of the sum, filling out its places with ciphers when it has not as many places as the repetends added; and carry the quotient to the next column.*

23. What is the sum of $3.\dot{6}+78.3\dot{4}7\dot{6}+735.\dot{3}+375.+.\dot{2}\dot{7}+187.\dot{4}$?

$3.\dot{6}$	$=\quad 3.666666\dot{6}$
$78.3\dot{4}7\dot{6}$	$=\; 78.347647\dot{6}$
$735.\dot{3}$	$=735.333333\dot{3}$
$375.$	$=375.000000\dot{0}$
$.\dot{2}\dot{7}$	$=\quad\;\; .272727\dot{2}$
$187.\dot{4}$	$=187.444444\dot{4}$
	$1380.0\dot{6}4819\dot{3}$

The sum of the repetends is first found to be 2648191. This sum is then divided by 999999, and it gives a quotient of 2, which we carry to the column of tenths, and a remainder of 648193, which we denote as a repetend.

24. What is the sum of $5391.35\dot{7}+75.3\dot{8}+187.\dot{2}\dot{1}+4.296\dot{5}+217.849\dot{6}+42.17\dot{6}+.5\dot{2}\dot{3}+58.3004\dot{8}$?

25. What is the sum of $9.8\dot{1}\dot{4}+1.\dot{3}+87.26+.08\dot{3}$ $+124.0\dot{9}$?

26. What is the sum of $.1\dot{6}\dot{2}+134.0\dot{9}+2.9\dot{3}+$ $97.2\dot{6}+3.\dot{7}6923\dot{0}+99.08\dot{3}+1.5+.\dot{8}1\dot{4}$?

SUBTRACTION OF INFINITE DECIMALS.

RULE. *Make the repetends similar and conterminous, and subtract as usual; observing, that, if the repetend of the subtrahend be greater than that of the minuend, the right hand figure of the remainder must be less by 1, than it would be, if the expression were finite.*

27. Subtract $13.7\dot{6}43\dot{2}$ from $85.\dot{6}\dot{2}$.

$$85.\dot{6}\dot{2} \quad=85.6\dot{2}626\dot{6}$$
$$13.7\dot{6}43\dot{2}=13.7\dot{6}643\dot{2}$$
$$\overline{\quad71.8\dot{6}19\dot{3}\quad}$$

Here, the whole repetend of the subtrahend is greater than that of the minuend, and the last figure in the remainder is diminished by 1.

28. Subtract $84.769\dot{7}$ from $476.3\dot{2}$.

29. Subtract $.03\dot{8}\dot{2}$ from $3.85\dot{6}\dot{4}$.

30. Subtract $493.1\dot{5}0\dot{2}$ from $1900.84\dot{2}97\dot{4}$.

MULTIPLICATION OF INFINITE DECIMALS.

RULE. *Change the factors to vulgar fractions, multiply these fractions together, and reduce their product to a decimal.*

31. What is the product of $.\dot{3}\dot{6}\times.2\dot{5}$?

$.\dot{3}\dot{6}=\frac{16}{99}=\frac{4}{11}$

$.2\dot{5}=\frac{23}{90}$ $\qquad \frac{4}{11}\times\frac{23}{90}=\frac{92}{990}=.09\dot{2}\dot{9}$ *Ans.*

32. What is the product of $27.2\dot{3}\times.2\dot{6}$?

33. What is the product of $8574.\dot{3}\times87.\dot{5}$?

34. What is the product of $3.9\dot{7}\dot{3}\times8$?

35. What is the product of $49640.54\times.\dot{7}050\dot{3}$?

36. What is the product of $3.1\dot{4}\dot{5}\times4.\dot{2}9\dot{7}$?

37. What is the product of $8.\dot{3}\times4.\dot{6}\times7.0\dot{9}$?

38. What is the product of $.\dot{3}\times.0\dot{9}\times8.\dot{2}\times.\dot{9}$?

DIVISION OF INFINITE DECIMALS.

RULE. *Change both divisor and dividend into vulgar fractions, find their quotient in a vulgar fraction, and reduce it to a decimal.*

39. What is the quotient of $.3\dot{6}$ by $.2\dot{5}$?

$.3\dot{6} = \frac{14}{99} = \frac{4}{11}$

$.2\dot{5} = \frac{23}{90}$ $\frac{4}{11} \div \frac{23}{90} = \frac{4}{11} \times \frac{90}{23} = \frac{360}{253} = 1\frac{107}{253}$

Then, $1\frac{107}{253} = 1.4229249011857707509881$ *Ans.*

40. What is the quotient of $234.\dot{6}$ by $.\dot{7}$?

41. What is the quotient of $13.5\dot{1}6953\dot{3}$ by $4.2\dot{9}\dot{7}$?

42. Divide $319.2800711\dot{2}$ by 764.5.

XIII.

RELATIONS OF NUMBERS.

Any number may be viewed as a part, or as so many parts of any other number; and it is in this view, that we shall, at present, notice the RELATION of one number to another.

For example, 1 is $\frac{1}{5}$ of 5, 3 is $\frac{3}{5}$ of 5, 9 is $\frac{9}{5}$ of 5, &c. Here 1 stands in the same relation to 5 that $\frac{1}{5}$ does to a unit; 3 stands in the same relation to 5 that $\frac{3}{5}$ does to a unit; and 9 stands in the same relation to 5 that $\frac{9}{5}$ does to a unit. Thus, the number which is viewed as the *part* or *parts* of another, becomes a numerator, and the other number the denominator of a vulgar fraction. This fraction may be reduced, and the relation it expresses will remain unaltered. For instance, $\frac{4}{8}$ of 8 is the same as $\frac{1}{2}$ of 8; and $\frac{24}{8}$ of 8 is the same as $\frac{6}{2}$ or $\frac{3}{1}$ of 8.

In the various practice of arithmetic, most of the solutions are performed by process to which the performer is led, by considering the relation which exists between the numbers concerned. The truth of this remark will appear evident to the learner, in the course of subsequent exercises.

1. Express 16 as a fractional part of 56, and reduce the fraction to its lowest terms.

2. Express 9 as the fractional part of 45, and reduce the fraction to its lowest terms.

3. What part of 34 is 20 ? What part of 34 is 21 ?

4. What part of 34 is 49 ?—Or, in other words, what is the improper fraction that expresses the relation in which 49 stands to 34 ?

5. What part of 24 is 36 ? What part of 24 is 37 ?

6. What part of 2 yards 1ft. 6in. is 1yd. 2ft. 10in.?

In this example, 2yd. 1ft. 6in. becomes a denominator, and 1yd. 2ft. 10in. the numerator. But both these quantities must be reduced to their lowest denomination, inches; the relation will then be simple, and may admit of being reduced to lower terms.

7. What part of 1 yard is 2 feet 6 inches ?

8. What part of £3 14s. is 16s. 10d.?

9. What part of 9s. 7d. 2qr. is 2s. 9d. 1qr.?

10. What part of 5 gallons 2 pints is 3 quarts 3 gills ?

11. What part of 2 acres is 1 acre 3 roods 32 rods ?

12. What part of $7 is $4.65 ?

$7 = 700 cents, and $4.65 = 465 cents. Then 465 cents is $\frac{465}{700}$ of 700 cents. $\frac{465}{700} = \frac{93}{140}$.·

When either or both the numbers, whose relation is to be expressed, contains a decimal fraction, the decimal places in the two numbers must be made equal—if they are not already so—by annexing decimal ciphers. The decimal points may then be erased, and the numbers written as the terms of a vulgar fraction. For example, the relation of .14 to 9 is $\frac{14}{900} = \frac{7}{450}$.

13. What part of 2.1 is 1.72 ?

14. What part of 4.87 is 2 ?

15. What part of $24.08 is $15 ?

16. What part of .65 is .408 ?

17. What part of $2 is $7 ? (*Ans.* $\frac{7}{2}$.)

18. What part of $2 is $7.49 ?

19. What part of 90 cents is $1.35 ?

20. What part of $4.375 is $28 ?

21. What part of 5.8 is 31.42?
22. What part of .253 is .97?

23. What part of $\frac{4}{7}$ is $\frac{6}{13}$?
The expression of this relation is, at first, a complex fraction, of which $\frac{6}{13}$ is the numerator, and $\frac{4}{7}$ the denominator. The expression may be simplified by reducing these fractions to a common denominator, and taking the new numerators for the terms of the relation. See rule, to reduce a complex fraction to a simple one, page 43.

24. What part of 12 is $10\frac{7}{9}$?
25. What part of 3 is $\frac{6}{11}$?
26. What part of $\frac{5}{7}$ is $\frac{4}{9}$?
27. What part of $6\frac{1}{4}$ is $5\frac{7}{8}$?
28. What part of $\frac{18}{34}$ is $\frac{22}{50}$?
29. What part of $2\frac{1}{3}$ feet is $10\frac{7}{8}$ inches?
30. What part of $14\frac{5}{6}$ days is $23\frac{7}{12}$ hours?
31. What part of 24 gallons is 3 quarts $2\frac{1}{4}$ gills?
32. What part of $5\frac{1}{4}$ rods is 3 rods $2\frac{2}{3}$ ft.?
33. What part of $\frac{1}{4}$ is $\frac{3}{4}$?
34. What part of $\frac{1}{4}$ is $\frac{1}{8}$?
35. What part of $\frac{2}{7}$ is $\frac{8}{9}$?
36. What part of $\frac{13}{24}$ is $\frac{36}{40}$?
37. What part of $1\frac{2}{3}$ is $3\frac{7}{8}$?
38. What part of 3 shillings is 5s. 7d.?
39. What part of £1 14s. is £5 2s. $7\frac{2}{3}$d.?
40. What part of $78\frac{1}{2}$ days is 125 days $17\frac{5}{9}$ hours?
41. What part of $2\frac{4}{7}$ tons is 4 tons $6\frac{1}{4}$ pounds?

42. If 35 horses eat 12278 pounds of hay in a week, what will 17 horses eat, in the same time?
The most obvious view of the solution of this question is this—If 35 horses eat 12278 pounds, 1 horse will eat $\frac{1}{35}$ of 12278 pounds, which is $350\frac{28}{35}$ pounds; and 17 horses will eat 17 times $350\frac{28}{35}$ pounds, which is 6230 pounds. A more concise view, however, may be taken, as follows. 17 horses are $\frac{17}{35}$ of 35 horses, and they will eat $\frac{17}{35}$ of the 12278 pounds of hay. Therefore, we shall obtain the answer by multiplying 12278 pounds by the fraction $\frac{17}{35}$. $12278 \times \frac{17}{35} = 6230$ *Ans.*

43. If a car run 552 miles upon a rail-road, in 24 hours, how far will it run in 13 hours?

44. If a car run 3 miles [960 rods] in 8 minutes [480 seconds], in what time will it run 300 rods?

45. If a hogshead of wine [63 gallons] cost $98.50, what will 45 gallons cost, at the same rate?

46. If the annual expense of supporting a fort manned with 600 soldiers be $182571, what is the expense of a fort manned with 424 soldiers?

47. If I can buy 325 barrels of flour for $1425, how many barrels can I buy for $521?

48. If a ferry boat cross the river 18 times in 5 hours, in how many hours will it cross 4 times?

49. If 9 barrels of flour cost $32, what will 28 bl. cost?

In this example, the relation in which 28 barrels stand to 9 barrels is expressed by an improper fraction; 28 barrels being $\frac{28}{9}$ of 9 barrels. Therefore the answer is obtained by multiplying $32 by $\frac{28}{9}$; that is, by multiplying $32 by 28, and dividing the product by 9.

50. If it take 300 yards of cloth to make the uniform clothes for 52 soldiers, how many yards are required to clothe 784 soldiers?

51. If 12 horses eat 20 bushels of oats in a week, how many bushels will 45 horses eat in the same time?

52. If a post 5 feet high cast a shadow 3 feet, on level ground, what is the height of a steeple, which, at the same time, casts a shadow 176 feet?

53. If $40 will pay for $14\frac{1}{2}$ yards of cloth, how many yards can be bought for $75?

54. If 95 bushels of corn cost $68.25, what will 320 bushels cost, at the same rate?

55. Suppose a ship's expenses in Liverpool to be £131 13s. 10d. for 22 days; what would be her expenses in the same port for 35 days?

56. If 144 bushels of corn will grow upon 3 acres 1 rood 15 rods of land, how much land is necessary to produce 500 bushels?

57. Bought 269 yards of cloth, at the rate of $100 for 30 yards. What did it amount to?

58. Bought 24 yd. 3 qr. 1 na. of cloth, at the rate of $12.30 for 4 yd. 1 qr. 2 na. What did it amount to?

Since it is necessary, in this example, to consider 24 yd. 3 qr. 1 na. as a fractional part of 4 yd. 1 qr. 2 na., the first step in the operation is, to reduce both quantities of cloth to nails.

59. If 13 gal. 2 qt. 1 pt. of wine cost $21.15, what will 36 gal. 3 qt. 1 pt. cost, at the same rate?

60. If 26 barrels of flour cost £28 14 s. 6 d. how many barrels will £35 10 s. 4 d. pay for?

61. If 6 gal. 2 qt. 1 pt. of wine will fill 31 bottles, how many bottles are required for 11 gal. 3 qt.?

62. If 144 gross of buttons cost £22 19 s., how many gross can be bought for £12 5 s. 5¼ d.?

63. If 2 hhd. 19 gal. 2 qt. of wine cost £93 1 s. 2¼ d. what will 25 hhd. 36 gal. cost?

64. If 15 yards of cloth cost $39.45, how many yards can be bought for $21? (See remark under example 12.)

65. At the rate of $94 for 78 days' work, in how many days can a labourer earn $72.375?

66. At the rate of $240 for 9.5 acres of land, what is the value of 7.25 acres?

67. At the rate of $182.50 for 8 acres of land, what is the value of 12.7 acres?

68. At the rate of 75 cents for .92 of a bushel of corn, what is the value of .648 of a bushel?

69. In how many minutes will a locomotive car run 49.9 miles; allowing it to run at the rate of 2.5 miles in 5.75 minutes?

70. If 43.64 pounds of copper be worth $9.075, what is the value of 108.9 pounds?

71. If 14 dollars will pay for the carriage of a ton 75.6 miles, what distance can a ton be carried for 16 dollars 75 cents, at the same rate?

72. If $\frac{4}{5}$ of a yard of cloth cost $7, what is the cost of $\frac{6}{13}$ of a yard? (Recur to example 23.)

73. If a rail-road car run 260 miles in 12 hours, what distance will it run in 10⅗ hours? (See example 24.)

74. If a man earn $1.15 in $\frac{5}{7}$ of a day, how much can he earn in $\frac{4}{9}$ of a day? (See example 25.)

75. Suppose $\frac{3}{4}$ of an acre of land to be worth 54 dollars; what is $\frac{1}{9}$ of an acre worth?

To solve this question, by the process to which the scholar has been led, he will consider $\frac{3}{4}$ as a denominator and $\frac{1}{9}$ as the numerator of a complex fraction, expressing what part of 54 dollars $\frac{1}{9}$ of an acre is worth; and, after reducing this complex fraction to a simple one, will multiply the simple fraction into 54 dollars, for the answer. Now the effect of the process is the same as that of multiplying the 54 by $\frac{1}{9}$, and dividing the product by $\frac{3}{4}$; and this last method is to be preferred, because it is shorter. Thus, $54 \times \frac{1}{9} = 6$, and $6 \div \frac{3}{4} = 8$.

76. If $\frac{7}{8}$ of a ship cost $15000, what does $\frac{1}{5}$ of her cost?

77. If $\frac{1}{3}$ of a lot of new land be worth 300 dollars, what is $\frac{7}{10}$ of the lot worth?

78. If a horse trot 1840 rods in $\frac{12}{13}$ of an hour, how many rods does he trot in $\frac{5}{18}$ of an hour?

79. If $96\frac{1}{4}$ yards of cloth cost $642, what will $28\frac{9}{16}$ yards cost, at the same rate?

80. If $15\frac{1}{2}$ yards of cloth cost $75, what will $142\frac{7}{8}$ cost, at the same rate?

81. If $9\frac{2}{3}$ barrels of flour be consumed by a company in 18 days, how long will $25\frac{1}{4}$ barrels last?

82. If a mill grind $18\frac{5}{16}$ bushels of corn in 1 hour and 22 minutes, in what time will it grind $25\frac{3}{4}$ bushels?

83. If a ship sail $92\frac{2}{3}$ miles in $8\frac{3}{4}$ hours, in how many hours does it sail $65\frac{3}{10}$ miles?

84. If a barrel of flour will support 12 men for 25 days, how long will it support 8 men?

Since the flour will support 12 men 25 days, it would support 1 man 12 times 25 days, or 300 days; and since it would support 1 man 300 days, it will support 8 men $\frac{1}{8}$ of 300 days, or $37\frac{4}{8}$ days. Thus, to obtain the answer, we multiply 25 days by 12, and divide the product by 8. A little attention to the conditions of this question, and the process of the operation, will enable the learner to perceive, at once, that the answer is $\frac{12}{8}$ of 25 days.

85. If a quantity of beef will support 436 men 73 days, how long will it support 240 men?

86. If a barrel of beer will last 10 men 16 days, how long will it last 23 men?

The beer would last 1 man 10 times 16 days, or 60 days; and it will last 23 men $\frac{1}{23}$ of 60 days, or $2\frac{14}{60}$ days $=2\frac{7}{30}$ days. The question is, however, more conveniently viewed thus,—Since the beer will last 10 men 16 days, it will last 23 men $\frac{10}{23}$ of 16 days; and, hence, 16 is to be multiplied by $\frac{10}{23}$.

87. Suppose a certain quantity of hay will feed 85 sheep 71 days; how long will it feed 230 sheep?

88. If 256 men can make a certain piece of road in 240 days, in what time will 190 men make it?

89. If 9 yards of silk, that is 3 quarters wide, will line a cloak, how many yards, that is 5 quarters wide, will line the same cloak?

90. If 110 yards of paper, that is 32 inches wide, will cover the walls of a room, how many yards, that is 24 inches wide, will cover the same walls?

91. Suppose a man can perform a piece of work in 45 days, by working 7 hours a day, in what time will he perform it, if they work 10 hours a day?

92. Suppose a company of men can perform a piece of work in 155 days, by working 12 hours a day, in what time will they perform it, by working 5 hours a day?

93. How many days will it take 119 horses to eat the hay that 44 horses would eat in 60 days?

94. The hind wheels of a coach, which are 180 inches in circumference, will turn round 4825 times in running a certain distance, how many times will the forward wheels turn round, they being 145 inches in circumference?

95. If a ship, by sailing 9 miles an hour, will effect a passage to Europe in 55 days, in how many days would she effect the passage by sailing 13 miles an hour?

96. If a vessel, by sailing $10\frac{1}{2}$ miles an hour, will make a passage from Bangor to New Orleans in 11 days, in how many days would she make the passage by sailing $12\frac{1}{4}$ miles an hour?

97. Suppose A rides $6\frac{7}{8}$ miles an hour, and performs

a certain journey in $14\frac{5}{12}$ days; in what time will B, who rides only $4\frac{9}{10}$ miles an hour, perform the same journey?

98. If 6 persons expend $300 in 8 months, how much will serve 15 persons for 20 months ? -

Since 6 persons expend $300 in 8 months, 15 persons would, in the same time, expend $\frac{15}{6}$ of $300, which is $750. Then, since 15 persons would expend $750 in 8 months, they would, in 20 months, expend $\frac{20}{8}$ of $750, which is $1875. The adjoined operation corresponds to this solution.

$$\begin{array}{cc} 300 & 750 \\ 15 & 20 \\ \hline 6)\overline{4500} & 8)\overline{15000} \\ \hline .750 & 1875 \end{array}$$

99. If the wages of 6 men for 14 days be $84, what will be the wages of 9 men for 11 days ?

100. If 3 pounds of yarn make 9 yards of cloth, 5 quarters wide, how many pounds would be required to make a piece of cloth 45 yd. long and 4 qr. wide?

101. If a class of 25 girls perform 1750 examples in arithmetic, in 15 hours, how many examples of equal length may a class of 30 girls perform, in 18 hours ?

102. If the use of $100 for 90 days, be worth $1.50, what is the use of $78 for 85 days worth ?

103. If the use of $100 for 30 days be worth 75 cents, what is the use of $1240 for 57 days worth ?

104. If a man travel 217 miles in 7 days, travelling 6 hours a day, how many miles will he travel in 9 days, if he travel 11 hours a day?

When he travels 6 hours a day, he advances 217 miles in 7 days, and were he to proceed thus for 9 days, he would advance $\frac{9}{7}$ of 217 miles, or 279 miles. Since, by travelling 6 hours a day he would, in 9 days, advance 279 ml., by travelling 11 hours a day, he would advance $\frac{11}{6}$ of 279 ml., which is $211\frac{3}{6}$ ml., or $211\frac{1}{2}$ ml.

105. If a man perform a journey of 1250 miles in 15 days, by travelling 14 hours a day, how many days will it take him, to perform a journey of 1000 miles, by travelling 13 hours a day?

106. If 10 cows eat $7\frac{1}{2}$ tons of hay in 14 weeks, how many cows will eat $22\frac{1}{2}$ tons in 28 weeks ?

107. If 6 men will mow 35 acres of grass in 7 days, by working 10 hours a day, how many men will be required to mow 48 acres in 5 days, when they work 12 hours a day?

108. If 14 men can cut 87 cords of wood in 3 days, when the days are 14 hours long, how many men will cut 175 cords, when the days are 11 hours long?

109. If 16 men can build 18 rods of wall in 12 days, how many men must be employed to build 72 rods of the same kind of wall in 8 days?

110. If 25 persons consume 600 bushels of corn in 2 years, how much will 139 persons consume in 7 years?

Since 25 persons consume 600 bushels in 2 years, 139 persons would, in the same time, consume $\frac{139}{25}$ of 600 bushels, which is 3336 bushels. Then, since 139 persons would consume 3336 bushels in 2 years, they will, in 7 years, consume $\frac{7}{2}$ of 3336 bushels, which is 11676 bushels.

111. If 154 bushels of oats will serve 14 horses for 14 days, how long will 406 bushels serve 7 horses?

112. If 25 men can earn $6250 in 2 years, how long will it take 5 men to earn $11250?

113. If 9 men can mow 36 acres of grass in 4 days, how many acres will 19 men mow in 11 days?

114. If a family of 9 persons spend $450 in 5 months, how much would be sufficient to maintain the family 8 months, after 5 more persons were added?

115. If a stream of water running into a pond of 190 acres, will raise the pond 10 inches in 12 hours, how much would a pond of 50 acres be raised by the same stream, in 10 hours?

116. If the wages of 4 men, for 3 days, be $11.04, how many men may be hired 16 days for $103.04?

117. If 3 men receive £8 18s. for working 19½ days what must 20 men receive for working 100¼ days?

118. If 1112 bottles are sufficient to receive 5 casks of wine, how many bottles are sufficient to receive 13 casks of wine?

119. If 725 bottles hold 4 barrels of wine, how many bottles are required to hold 3 tierces of wine?

120. If 240 men, in 5 days, of 11 hours each, can dig a trench 230 yards long, 3 yards wide, and 2 yards deep, in how many days, of 9 hours each, will 24 men dig a trench 420 yards long, 5 yards wide, and 3 yards deep?

Since 248 men, in 5 days, of 11 hours each, can dig a trench 230 yards long, 3 yards wide, and 2 yards deep, 24 men, working in days of the same length, would dig a trench of the same dimensions in $\frac{248}{24}$ of 5 days, which is $51\frac{1}{2}\frac{1}{3}=51\frac{2}{3}$ days; and, working in days of 9, instead of 11 hours each, the trench would occupy them $\frac{11}{9}$ of $51\frac{2}{3}$ days, which is $63\frac{4}{27}$ days. Again, since the trench to be dug by 24 men is 420, instead of 230 yards long, this length, (the width and depth remaining unchanged) would occupy them $\frac{420}{230}=\frac{42}{23}$ of $63\frac{4}{27}$ days, which is $115\frac{65}{207}$ days. Again, since the trench to be dug by 24 men is 5, instead of 3 yards wide, this width (the depth remaining unchanged) would occupy them $\frac{5}{3}$ of $115\frac{65}{207}$ days, which is $192\frac{118}{621}$ days. Lastly, since the trench to be dug by 24 men is 3, instead of 2 yards deep, it will occupy them $\frac{3}{2}$ of $192\frac{118}{621}$ days, which is $288\frac{177}{621}=288\frac{59}{207}$ days, the answer.

121. If 12 men can build a brick wall 25 feet long, 7 feet high, and 4 feet thick, in 18 days, in how many days will 20 men build a brick wall 150 feet long, 8 feet high, and 5 feet thick?

122. If 15 men can dig a trench 75 feet long, 8 ft. wide, and 6 ft. deep, in 12 days, how many men must be employed to dig a trench 300 ft. long, 12 ft. wide, and 9 ft. deep, in 10 days?

123. If the carriage of 44 barrels of flour, 108 miles be worth $215, what is the carriage of 36 barrels, 162 miles worth?

124. If 175 bushels of corn, when corn is worth 60 cents a bushel, be given for the carriage of 100 barrels of flour, 58 miles, how many bushels of corn, when corn is worth 75 cents a bushel, must be given for the carriage of 90 barrels of flour, 200 miles?

125. If 12 ounces of wool make $2\frac{1}{2}$ yards of cloth, that is 6 quarters wide, how many pounds of wool would make 150 yards of cloth, 4 quarters wide?

MISCELLANEOUS EXAMPLES.

126. A owned $\frac{5}{24}$ of a ship, which he sold for $\$3650$, and B owns $\frac{3}{10}$ of her, which he wishes to sell at the same rate. What must be B's price?

Since the price of $\frac{5}{24}$ of the ship is $\$3650$, the price of the whole ship must be $\frac{24}{5}$ of $\$3650$, which is $\$36900$; and $\frac{3}{10}$ of $\$36900$ is $\$11070$, which must be B's price.

127. If 3650 be $\frac{5}{24}$ of some number, what is $\frac{3}{10}$ of the same number?

128. A merchant has bought $\frac{7}{18}$ of a company's stock, for $\$92000$. What would be the price of $\frac{4}{23}$ of the stock, at the same rate?

129. A merchant owning $\frac{7}{12}$ of a ship, sold $\frac{9}{32}$ of what he owned for $\$1841$. What is the value of the whole ship, according to this sale?

130. 1841 is $\frac{9}{32}$ of $\frac{7}{12}$ of what number?

131. After a certain tract of land had been equally divided among 16 owners, one of them sold $\frac{2}{5}$ of his share at $\$5$, an acre, and received $\$444$. How much land was there in the whole tract?

132. If $\frac{7}{16}$ of a yard of cloth be worth $\frac{5}{6}$ of a dollar, what is the value of $\frac{9}{11}$ of a yard?

Since $\frac{7}{16}$ of a yard is worth $\frac{5}{6}$ of a dollar, a yard is worth $\frac{16}{7}$ of $\frac{5}{6}$ of a dollar, which is $\frac{80}{42}$ of a dollar; and $\frac{9}{11}$ of a yard is worth $\frac{9}{11}$ of $\frac{80}{42}$ of a dollar, which is $\frac{720}{462}$ of a dollar, or $\$1\frac{258}{462} = \$1\frac{43}{77} = \$1.559+$.

133. If $\frac{3}{8}$ of a yard of lace be worth $\frac{17}{18}$ of a dollar, what is $\frac{9}{13}$ of a yard worth?

134. If $\frac{3}{4}$ of a barrel of flour cost 4 dollars, what is the cost of $6\frac{2}{3}$ barrels, at the same rate?

135. If $13\frac{7}{8}$ bushels of corn cost 7 dollars, what is the price of $9\frac{1}{2}$ bushels, at the same rate?

136. If $42\frac{3}{4}$ pounds of indigo be worth $\$87.625$, what is the value of $192\frac{3}{8}$ pounds?

137. A garrison of 900 men have provision for 4 months. How many men must leave the garrison, that the provision may last the remainder 9 months?

138. If a loaf of bread weighing 32 ounces be sold for eight cents, when flour is worth $\$6.50$ per barrel, what

ought the eight-cent loaf to weigh, when flour is worth only $5 a barrel?

139. A company of 75 soldiers are to be clothed; each suit is to contain 3¼ yards of cloth, 6 quarters wide, and to be lined with flannel ⅝ of a yard wide. How many yards of flannel will be required?

140. If a garrison of 1500 men consume 750 barrels of flour in 9 months, how many barrels will 2150 men consume in 15 months?

141. How many tiles 8 inches square, will cover a hearth 16 feet long, and 12 feet wide?

142. If the expense of carrying 17 cwt. 3 qr. 14 lb. 85 miles be $23.84, what will be the expense of carrying 53 cwt. 2 qr. 150 miles, at the same rate?

143. Two men bought a barrel of flour; one paid 3½ dollars, and the other paid 3⅔ dollars. What part of the flour should each of them have?

144. If the corn contained in 8 bags, holding 2 bushels 3 pecks each, be worth $14.25, what is the value of the corn contained in 7 bags, each holding 2 bu. 3 pk. 7 qt.?

145. A ship of war sailed with 650 men, and provision for a cruise of 15 months. At the end of 3 months she captured an enemy's vessel, and put 75 men on board of her. Five months after, she captured and sunk another vessel, and took on board the crew, consisting of 350 men. How long did the provision last, from the commencement of the cruise?

146. A built 156 rods of wall in a certain time, and B in the same time built 13 rods to every 12 that A built. They were paid $1.25 per rod. How much did B receive more than A?

147. A father bequeathed $6000 as follows; viz. ⅜ to his wife, ½ to his son, ¼ to his daughter, and the remainder to his servant. How much did each receive?

148. If 11/12 of a pound of sugar be worth ⅓ of a shilling, what is the value of 7/8 of a cwt.?

149. If 75 7/15 gallons of water, in one hour, run into a cistern, which will hold 6½ hogsheads, and by a pipe 24¼ gallons an hour run out, in how many hours, minutes and seconds will the cistern be filled?

8

XIV.

PERCENTAGE.

Under this head may be classed, those computations which investigate the value of a given number of hundredths of any quantity. The number of hundredths to be taken or considered in any number, is called the *per cent.* The term, *per cent.*, is an abbreviation of *per centum*, which signifies *by the hundred.*

Any per cent. is conveniently expressed by a decimal. Thus, 1 per cent. of any number is .01 of that number; 8 per cent. is .08; 25 per cent. is .25; &c.

1. A merchant, who has 426 dollars deposited in the bank, wishes to draw out 5 per cent. of his deposite. How many dollars must he draw?

Since 5 per cent. of any quantity is $\frac{5}{100}$ of that quantity, the question to be solved in this example is— What is $\frac{5}{100}$ of 1426 dollars? Or, decimally— What is .05 of 1426 dollars? The answer is conveniently found by multiplying 1426 by .05. The whole number in the product expresses dollars, and the decimal expresses cents.

$$\begin{array}{r} 1426 \\ .05 \\ \hline \$71.30 \end{array}$$

2. A trader, who went to the city with 321 dollars, to purchase goods, laid out 9 per cent. of his money for coffee. How many dollars did he pay for coffee?

3. What is 1 per cent. of 100 dollars?
4. What is 1 per cent. of 834 dollars?
5. What is 3 per cent. of 100 dollars?
6. What is 3 per cent. of 42 dollars?
7. What is 6 per cent. of 100 dollars?
8. What is 6 per cent. of 99 dollars?
9. What is 7 per cent. of 100 dollars?
10. What is 7 per cent. of 1000 dollars?
11. What is 8 per cent. of 26 dollars?
12. What is 9 per cent. of 354 dollars?
13. What is 10 per cent. of 2244 dollars?

14. What is 16 per cent. of 13 dollars?

15. What is 37 per cent. of 211 dollars?

16. What is 99 per cent. of 100 dollars?

17. What is 100 per cent. of 48 dollars?

18. A trader laid out 1214 dollars as follows. He paid 24 per cent. of the money for broadcloths; 38 per cent. for linens; 8 per cent. for calicoes; and the remainder for cottons. How many dollars did he pay for each kind of goods?

When the rate per cent. is a vulgar fraction, or a mixed number, the fraction may be changed to a decimal. Observe, that, 1 per cent. when expressed decimally, is .01; therefore a fraction of 1 per cent. when reduced to a decimal, becomes so many tenths, hundredths, &c. of a hundredth. For example, as $\frac{1}{4}$ of 1 unit is .25 of a unit, so $\frac{1}{4}$ of 1-hundredth is .25 of a hundredth, and is denoted thus, .0025.

19. What is $3\frac{1}{2}$ per cent. of 243 dollars?

$$\begin{array}{ll} .3 \text{ per cent.} = .03 & \quad 243 \\ \frac{1}{2} \text{ per cent.} = .005 & \quad .035 \\ \hline \quad\quad .035 & \quad 1215 \\ & \quad 729 \\ & \overline{\quad 8.505} \quad \textit{Ans.} \ \$8.50\frac{1}{2} \end{array}$$

20. What is $4\frac{1}{2}$ per cent. of 2746 dollars?

21. What is $7\frac{1}{2}$ per cent. of 41 dollars?

22. What is $12\frac{1}{2}$ per cent. of 358 dollars?

23. What is $\frac{1}{2}$ per cent. of 100 dollars?

24. What is $\frac{1}{2}$ per cent. of 61 dollars?

25. What is $\frac{1}{4}$ per cent. of 9487 dollars?

26. If $8\frac{1}{4}$ per cent. be taken from 36 dollars, how many dollars will there be remaining?

27. A merchant who had 400 barrels of flour, shipped $42\frac{1}{2}$ per cent. of it, and sold the remainder. How many barrels did he sell?

28. A trader bought 800 pounds of coffee; and, in getting it to his store, $2\frac{1}{4}$ per cent. of it was wasted. How many pounds did he lose? What did the remainder amount to, at 13 cents a pound?

29. Two men had 120 dollars each. One of them paid out 14 per cent. of his money, and the other $17\frac{1}{2}$ per cent. How many dollars did one pay more than the other?

30. Find $7\frac{1}{3}$ per cent. of $344.

When there is a fraction in the rate per cent. which cannot be exactly expressed by a decimal— as in this example— we first find 1 per cent. of the given sum, by dividing it by 100; that is, by cut-

$$344 \div 100 = 3.44$$
$$7\frac{1}{3}$$
$$\overline{2408}$$
$$114\frac{2}{3}$$
$$\$25.22\frac{2}{3}$$

ting off two decimal figures, and then multiply this quotient by the mixed number expressing the rate per cent.

31. What is $4\frac{1}{2}$ per cent. of 624 dollars?
32. What is $6\frac{2}{3}$ per cent. of 38 dollars?
33. What is $3\frac{1}{4}$ per cent. of 2310 dollars?
34. What is $9\frac{1}{6}$ per cent. of 17 dollars?
35. What is $8\frac{1}{4}$ per cent. of 152 dollars?

36. Find the difference between $5\frac{2}{3}$ per cent. of 41 dollars, and $4\frac{1}{3}$ per cent. of 39 dollars.

37. What is 7 per cent. of $24.32?

Here we have cents [decimals] in the number on which the percentage is to be taken. We however multiply as usual in decimal multiplication; and the first two decimal fig-

$$24.32$$
$$.07$$
$$\overline{\$1.7024}$$

ures in the product express cents, the third figure expresses mills, and the fourth expresses *tenths* of a mill.

38. What is 14 per cent. of $641.94?
39. What is $4\frac{1}{2}$ per cent. of $37.26?
40. What is $11\frac{1}{2}$ per cent. of $150.75?
41. What is $12\frac{1}{4}$ per cent. of $25.32?

42. If a horse and gig cost 400 dollars, and the gig cost 32 per cent. of the sum, what did the horse cost?

43. Find the difference between $13\frac{1}{2}$ per cent. of $18.09, and 7 per cent. of $41.

44. Find the difference between 9 per cent. of $16, and $8\frac{1}{4}$ per cent. of $17.30.

45. A young man, who had 94 dollars deposited in the Savings Bank, drew out 25 dollars. What per cent. of his deposite did he draw out?

We perceive, that the sum he drew out, was $\frac{25}{94}$ of the sum he had deposited: and, since the rate per cent. of any sum is a certain number of hundredths of that sum, the ques-

$$94)25.0(.26\tfrac{56}{94}=26\tfrac{24}{47}$$
$$188$$
$$\overline{620}$$
$$564$$
$$\overline{56}$$

tion to be solved is— How many hundredths is 25-ninety-fourths?— To solve this question, we change $\frac{25}{94}$ to a decimal; restricting the decimal to hundredths; that is, carrying the quotient no further than two places. Any remainder which might allow the quotient to be carried further, may, in cases like this, be expressed in a vulgar fraction.   *Ans.* $26\tfrac{24}{47}$ per cent.

46. A man, who was owing a debt of 240 dollars, has paid 32 dollars of it. What per cent. of the debt has he paid?

47. A merchant gave his note for 235 dollars, and soon after paid 110 dollars of the sum. What per cent. did he pay; and what per cent. still remained due?

48. If the cloth for a coat cost 12 dollars, and the making 7 dollars, what per cent. of the whole expense is the making?

49. What per cent. of 100 dollars is 6 dollars?.

50. What per cent. of $28.50 is $1.10?

51. What per cent. of $94.12 is $4.42?

52. What per cent. of $57.08 is 32 cents?

53. What per cent. of $10.10 is 7 cents?

54. What per cent. of $48.11 is 99 cents?

55. What per cent. of $75 is $4.18?

To find the value of a rate per cent. on any sum of English money,— First, change the lower denominations of money in the sum, to a decimal of the highest denomination; and then proceed to multiply by the rate, as if the sum were dollars and cents. The whole number in the product will be of the same denomination of money with the whole number in the multiplicand; and the

8 *

decimal in the product must be changed to the lower denominations.

56. An English gentleman took passage from Liverpool to Boston, in the ship Dover, having £672 12s. 4d. He paid 5 per cent. of his funds for his passage. How much did he pay?

```
12| 4.                  .6308
20|12.333+               20
   ─────────         ─────────
   672.616+·         12.6160
      .05                12
   ─────────         ─────────
   33.63080          7.3920
                         4
                     ─────────
              3.680 Ans. £33 12s. 7d. 3qr.+
```

57. What is 8 per cent. of £47 18s. 7d.?
58. What is 3 per cent. of £9 14s. 3qr.?
59. What is 16 per cent. of £22 16s.?
60. What is 25 per cent, of 19s. 8d. 2qr.?
61. What is 6 per cent. of £2584?
62. What is 50 per cent. of 18s. 10d. 2qr.?
63. What is 4½ per cent. of £214 15s. 10d.?

COMMISSION.

COMMISSION is the compensation made to factors and brokers for their services in buying or selling. It is reckoned at so much per cent. on the money employed in the transaction.

64. What is the commission on £500 at 2½ per cent.?
65. Suppose I allow my correspondent a commission of 2 per cent., what is his demand on the disbursement of £369?
66. If I allow my factor a commission of 3 per cent. for disbursing £748 11s. 8d. on my account, what does his commission amount to?
67. How much does a broker receive on a sale of stocks amounting to 52648 dollars, allowing his commission to be ¼ of 1 per cent.?

68. What is the amount of commission on 395 dollars 75 cents, at $3\frac{1}{2}$ per cent.?

69. A commission merchant sold goods to the amount of 6910 dollars and 80 cents, upon which he charged a commission of $2\frac{1}{2}$ per cent. How much money had he to pay over to his employer?

70. Sold 94 tons, 17 cwt. 3 qr. of iron, at 96 dollars a ton, at a commission of $2\frac{1}{2}$ per cent. on the sale. What did my commission amount to? How much had I to pay over?

STOCKS.

STOCK is a property, consisting in shares of some establishment, designed to yield an income. It includes government securites, shares in incorporated banks, insurance offices, factories, canals, rail-roads, &c.

The *nominal* value, or *par value* of a share, is what it originally cost; and the *real* value, at any time, is the sum for which it will sell. When it will sell for more than it originally cost, it is said to be *above par*, and the excess is stated at so much per cent. *advance*. When its real value is less than the original cost, it is said to be *below par*, and is sold at a discount.

71. Sold 10 shares in the Commonwealth Insurance Company, at 5 per cent. advance, the par value of a share being 100 dollars. How much did I receive?

72. Bought 15 shares in the Boston Bank, at $\frac{3}{4}$ of 1 per cent. advance, the par value being 50 dollars a share. How much did I give for them?

73. Sold 64 shares in the State Bank, at $1\frac{1}{4}$ per cent. advance, the par value being 60 dollars a share. How much did I receive for them?

74. Sold 9000 dollars United States 5 per cent. stock, at an advance of $7\frac{1}{4}$ per cent. What was the amount of the sale?

75. Sold 18 shares in an insurance office, at $1\frac{3}{8}$ per cent. discount, the par value being 100 dollars a share. How much did they come to?

76 Bought 16 shares in the Massachusetts Bank, at $1\frac{1}{4}$ per cent. advance, the par value being 250 dollars a share. What was the amount of the purchase?

77. Bought 54 shares in the New York City Bank, at $7\frac{3}{4}$ per cent. advance, the par value being 100 dollars a share. How much did they cost me?

78. I directed a broker to purchase 25 shares of railroad stock, at a discount of 13 per cent. the par value being $100 per share. Allowing the broker's commission to be $\frac{7}{8}$ per cent., what will the whole cost me?

79. What will 16 shares in the Philadelphia Bank cost; the par value being $100 per share, the price being $3\frac{1}{2}$ per cent. above par, and the broker charging a commission of $\frac{3}{4}$ per cent.?

INSURANCE.

INSURANCE is security given, to restore the value of ships, houses, goods, &c., which may be lost by the perils of the sea, or by fire, &c. The security is given in consideration of a *premium* paid by the owner of the property insured.

The premium is always a certain per cent. on the value of the property insured, and is paid at the time the insurance is effected.

The written instrument, which is the evidence of the contract of indemnity, is called a *policy.*

80. What is the amount of premium for insuring 19416 dollars at $2\frac{1}{2}$ per cent.?

81. I effected an insurance of 3460 dollars on my dwelling house for one year at $\frac{3}{8}$ of .1 per cent. What did the premium amount to?

82. If you obtain an insurance on your stock of goods valued at 7325 dollars, at $\frac{1}{2}$ of 1 per cent. what will the premium amount to?

83. If you should take out a policy of 3168 dollars, on your store and goods, at a premium of 41 cents on a hundred dollars, what would be the amount of premium?

84. An insurance of 18000 dollars was effected on

the ship Sturdy, on her last voyage from Boston to Calcutta, at a premium of 3 per cent. out and home. What did the premium amount to ?

85. An insurance of 3500 dollars on stock in a cotton factory was effected at $3\frac{1}{4}$ per cent. for one year. What was the amount of premium ?

86. A gentleman procured an insurance for one year on his house valued in the policy at 8756 dollars, and on his furniture valued at 2139 dollars, at a premium of 39 cents on a hundred dollars. How much did the premium amount to ?

XV.

INTEREST.

INTEREST is a premium paid for the use of money.

It is computed by percentage; a certain per cent. on the money being paid for its use, for a stated time.

The money on which interest is paid, is called the *Principal*. The per cent. paid, is called the *Rate*. The principal and interest added together, are called the *Amount*.

When a rate per cent. is stated without the mention of any term of time, the time is understood to be 1 year.

The rate of interest is regulated by state laws, and is not uniform in all the states. We shall, however, first treat of 6 per cent. per annum, as this is the rate most commonly paid.

As interest is always expressed by some rate per cent., the most convenient way of computing it is, to find the decimal expression of the rate for the time, and multiply the principal by this decimal: the product is the interest. Thus, if the rate for 1 year be 6 per cent. or .06, for 2 years it is 12 per cent. or .12, for 3 years it is 18 per cent. or .18, and so on. The interest of 24 dollars for 3 years, at 6 per cent. a year, is found thus, $24 \times .18 =$ $4.32, the interest sought.

RATE PER CENT. FOR MONTHS. The decimal expression of the rate for months, when the rate is 6 per cent. a year, is easily obtained; for, if the rate for 12 months be 6 per cent. or .06, for 1 month it is $\frac{1}{12}$ of 6 per cent. which is $\frac{1}{2}$ per cent. or .005; for 2 months it is 1 per cent. or .01; for 3 months it is $1\frac{1}{2}$ per cent. or .015; for 4 months it is 2 per cent. or .02; for 5 months it is $2\frac{1}{2}$ per cent. or .025; for a year and 1 month it is $6\frac{1}{2}$ per cent. or .065; for a year and 2 months it is 7 per cent. or .07; for a year and 11 months it is $11\frac{1}{2}$ per cent. or .115.

1. If the rate of interest be 6 per cent. for a year, what is the rate for 1 month? ...... for 6 months? ...... for 7 months? ...... for 8 months? ...... for 9 months?

2. At 6 per cent. a year, what is the rate for a year and 1 month? ...... a year and 3 months? ...... a year and 4 months? ...... a year and 10 months?

RATE PER CENT. FOR DAYS. Observe, that the rate for 2 months, which is 60 days, is 1 per cent. or .01; and for $\frac{1}{10}$ of 60 days, which is 6 days, it is $\frac{1}{10}$ of .01, which is .001. Now since the rate for 6 days is 1-thousandth, the rate for any number of days is as many thousandths as there are times 6 days. Therefore, to find the rate for days, at 6 per cent. per annum, adopt the following RULE. *Denote the days as so many thousandths, and divide the expression by 6: the quotient will be the rate.*

3. If the rate of interest be 6 per cent. for a year, what is the rate for 1 day? ...... for 2 days? ...... for 3 days? ...... for 4 days? ...... for 5 days? ...... for 6 days? ...... for 7 days? ...... for 9 days? ...... for 24 days? ...... for 26 days?

4. At 6 per cent. a year, what is the rate for 2 months and 12 days? ...... 3 months and 10 days? ...... for 5 months and 18 days? ...... for 10 months and 29 days?

5. What is the interest, and what the amount of 546 dollars 72 cents, for 4 years 7 months 19 days, at 6 per cent. a year?

To find the rate for 4 years, we multiply the rate for 1 year by 4; thus, .06 × 4 = .24. To find the rate for 7 months, we multiply the rate for 1 month by 7; thus, .005 × 7 = .035. To find the rate for 19 days, we denote 19 as *thousandths*, and divide the expression by 6; thus, .019 ÷ 6 = .00316 +. Now the sum of these rates, .24 + .035 + .00316 = .27816, is the rate for the whole time; and by this sum we multiply the principal.

```
        546.72
        .27816
        328032
         54672
        437376
        382704
        109344
      152.0756352
        546.72
      698.7956352
```

The interest found, is $ 152.07,5 +; which, added to the principal, gives the amount, $ 698 .79,5 +. The rate for 19 days is not exact, as the decimal does not terminate; it is, however, sufficiently near exactness.

6. What is the interest of 148 dollars 92 cents, for 3 years, at 6 per cent. per annum?

7. What is the interest of 57 dollars 10 cents, for 5 years, at 6 per cent. a year?

8. What is the interest of 93 dollars 50 cents, for 4 years, at 6 per cent. a year?

9. What is the interest of 608 dollars 62 cents, for a year and 9 months, at 6 per cent. a year?

10. What will 713 dollars 33 cents amount to, in 2 years and 10 months, at 6 per cent. per annum?

11. What will 1256 dollars 81 cents amount to, in 8 months, at the rate of 6 per cent. a year?

12. What is the interest of 100 dollars, for 1 year 11 months and 24 days, at 6 per cent. a year?

13. To what sum will 37 dollars 50 cents amount, in 1 year 7 months and 21 days, at 6 per cent. per annum?

14. What is the interest of 314 dollars 36 cents, for 1 year 1 month and 6 days, at 6 per cent. a year?

15. What is the interest of 37 dollars 87 cents, for 11 months and 15 days, at 6 per cent. a year?

16. What is the interest of 512 dollars 38 cents, for 7 months and 10 days, at 6 per cent. a year?

17. To what sum will 691 dollars 28 cents amount, in 1 year and 1 month, at 6 per cent. a year?

18. What is the amount of 194 dollars 69 cents, for 1 year 5 months and 6 days, at 6 per cent. a year? .

19. What will 32 dollars 47 cents amount to, in 9 months and 25 days, at 6 per cent. a year?

20. What is the interest of 217 dollars 19 cents, for 1 year and 17 days, at 6 per cent. a year?

21. What is the amount of 143 dollars 37 cents, for 1 year 9 months and 4 days, at 6 per cent. per annum?

22. To what sum will 203 dollars 9 cents amount, in 2 years and 19 days, at 6 per cent. per annum?

23. To what sum will 18 dollars 63 cents amount, in 1 year 10 months and 19 days, at 6 per cent. a year?

24. What is the interest of 600 dollars, for 7 months and 22 days, at 6 per cent. a year?

25. What is the interest of 817 dollars 44 cents, for 11 months and 12 days, at 6 per cent. a year?

26. What is the interest of 155 dollars, for 1 year 2 months and 10 days, at 6 per cent. a year?

27. To what sum will 109 dollars 12 cents amount, in 5 months and 8 days, at 6 per cent. a year?

28. What is the amount of 25 dollars 92 cents, for 1 year 4 months and 7 days, at 6 per cent. a year?

29. To what sum will 65 dollars 48 cents amount, in 1 year 1 month and 18 days, at 6 per cent. a year?

30. What is the interest of 110 dollars 25 cents, for 10 months and 4 days, at 6 per cent. a year?

31. What is the interest of 2814 dollars 70 cents, for 6 months and 3 days, at 6 per cent. a year?

32. What is the amount of 84 dollars 33 cents, for 8 months and 26 days, at 6 per cent. per annum?

33. What is the interest of 345 dollars 68 cents, for 7 months and 13 days, at 6 per cent. a year?

34. To what sum will 13 dollars 98 cents amount, in 2 years 4 months and 7 days, at 6 per cent. a year?

35. What is the interest of 802 dollars 27 cents, for 1 month and 5 days, at the rate of 6 per cent. a year?

36. What is the interest of 1309 dollars, for 2 months and 3 days, at the rate of 6 per cent. a year?

37. To what sum will 23 dollars 8 cents amount, in 3 years 6 months and 22 days, at 6 per cent. a year?

38. What is the interest of 2538 dollars 17 cents, for 3 months and 28 days, at the rate of 6 per cent. a year?

39. What is the amount of 1800 dollars 34 cents, for 1 year and 2 days, at 6 per cent. a year?

40. What is the interest of 199 dollars 15 cents, for 1 year and 23 days, at 6 per cent. a year?

41. To what sum will 49 dollars 5 cents amount, in 1 year 2 months and 3 days, at 6 per cent. a year?

42. What is the interest of 201 dollars 50 cents, for 7 years, at 6 per cent?

43. What is the interest of 3010 dollars 75 cents, for 3 months and 1 day, at the rate of 6 per cent. a year?

44. To what sum will 41 dollars 6 cents amount, in 1 year 5 months and 14 days, at 6 per cent. a year?

45. What is the amount of 50 dollars and 11 cents, for 1 year and 21 days, at 6 per cent. a year?

46. What is the interest of 1100 dollars for a year and 15 days, at 6 per cent. a year?

47. What is the interest of 9 dollars 89 cents, for 1 year and 27 days, at 6 per cent. a year?

48. What is the interest of 80 dollars, for 1 year 5 months and 12 days, at 6 per cent a year?

49. What is the interest of 90 dollars, for 1 year 2 months and 6 days, at 6 per cent. a year?

50. To what sum will 55 dollars amount, in 3 years and 9 days, at 6 per cent. a year?

51. What is the amount of 4119 dollars 20 cents, for 1 year and 5 days, at 6 per cent. a year?

To compute interest by DAYS, when the rate is 6 per cent. per annum. RULE. *Multiply the principal by the number of days, and divide the product by 6. The quotient is the interest in mills, when the principal consists of dollars only; but when there are cents in the principal, cut off two figures from the right of the quotient, and the remaining figures will express the mills.*

This rule—like the rule for finding the *per cent.* for days— is based upon the supposition of 360 days to the year; and, since the year contains 365 days, the rule gives $\frac{1}{73}$ part more than a true six per cent. interest.

9

52. What is the interest of 86 dollars, for 20 days, at 6 per cent. a year?

53. What is the amount of 108 dollars, for 25 days, at 6 per cent. a year?

54. What is the interest of 204 dollars, for 40 days, at 6 per cent. a year?

55. What is the interest of 1000 dollars, for 29 days, at 6 per cent. a year?

56. What is the amount of 98 dollars 60 cents, for 35 days, at 6 per cent. a year?

57. What is the interest of 250 dollars, for 18 days, at 6 per cent. a year?

58. What is the interest of 61 dollars 25 cents, for 28 days, at 6 per cent. a year?

59. What is the amount of 215 dollars 78 cents, for 50 days, at 6 per cent. a year?

60. What is the interest of 71 dollars, for 41 days, at 6 per cent. a year?

61. What is the interest of 3333 dollars, for 10 days, at 6 per cent. a year?

62. What is the amount of 37 dollars 58 cents, for 16 days, at 6 per cent. a year?

63. What is the interest of 91 dollars 80 cents, for 57 days, at 6 per cent. a year?

64. What is the interest of 4109 dollars, for 18 days, at 6 per cent. a year?

65. What is the amount of 5214 dollars, for 50 days, at 6 per cent. a year?

66. What is the difference between the interest of $1000 for 1 year, computed by the year, and the interest on the same sum for the same time, computed by days; both at 6 per cent.?

It will be observed, that, in all the preceding examples, the rate of interest has been 6 per cent. per annum. The method of computing interest at any other rate per cent. is the same, and equally simple, when the time consists of years only; but when there are months and days in the time, and the rate per cent. per annum is other than 6, it will frequently be convenient to find the interest for a year first; and then for the months, to take the aliquot parts of a year; and for the days, the aliquot parts of a month; as in the following examples.

67. What is the interest of 934 dollars 34 cents, for 3 years and 5 months, at 7 per cent. per annum ?

$$934.34$$
$$.07$$

4 months is ⅓ of a Y. 3) 65.4038   interest for 1 year
$$3$$

196.2114   interest for 3 years.
1 month is ¼ of 4 ms. 4) 21.8012   interest for 4 ms.
5.4503   interest for 1 m.

$ 223.4629   interest for 3 Y. 5 ms.

68. What is the interest of 371 dollars 42 cents, for 1 year 9 months and 19 days, at 7½ per cent. a year ?

$$371.52$$
$$.075$$

185760
260064

6 months is ½ of a year. 2) 27.86400   for 1 year.
3 months is ½ of 6 ms. 2) 13.93200   for 6 months.
15 days is ⅕ of 3 ms. 6) 6.96600   for 3 months.
3 days is ⅕ of 15 days. 5) 1.16100   for 15 days.
1 day is ⅓ of 3 days 3) .23220   for 3 days.
.07740   for 1 day.

$ 50.53268   for the whole time

69. What is the interest of 412 dollars 17 cents, for 1 year 7 months and 10 days, at 7 per cent. a year ?

70. What is the interest of 15748 dollars, for a year, at 4½ per cent.?

71. What is the interest of 125 dollars 50 cents, for 2 years at 7 per cent. a year ?

72. What is the interest of 969 dollars, for 4 years, at 8 per cent. a year ?

73. What is the interest of 655 dollars 30 cents, for a year, at 7 per cent.?

74. What is the interest of 404 dollars 39 cents, for a year, at 5½ per cent.?

75. To what sum will 1060 dollars 90 cents amount, in a year, at 7 per cent.?

76. What is the interest of 1650 dollars, for a year, at 30 per cent.?

77. What will 1428 dollars amount to, in a year and 5 months, at 5 per cent. a year?

78. What is the interest of 2194 dollars 50 cents, for a year and 10 months, at 7 per cent. a year?

79. What is the interest of 20750 dollars 42 cents, for 1 year 2 months and 20 days, at 4½ per cent. a year?

80. What is the interest of 1109 dollars 44 cents, for 11 months, at 5½ per cent. a year?

81. What is the interest of 717 dollars 19 cents, for 5 months and 6 days, at the rate of 7 per cent. a year?

82. What is the interest of 2119 dollars 78 cents, for 3 months and 24 days, at 4½ per cent. a year?

83. To what sum will 107 dollars 29 cents amount, in 7 months and 5 days, at the rate of 7 per cent. a year?

84. To what sum will 5128 dollars 60 cents amount, in 3 months and 26 days, at 5½ per cent. a year?

85. What is the interest of 8244 dollars, for 1 month and 20 days, at the rate of 8 per cent. per annum?

86. What is the interest of 1062 dollars 80 cents, for 2 months, at the rate of 9 per cent. per annum?

87. What is the interest of 4008 dollars 90 cents, for 9 months, at the rate of 7½ per cent. a year?

88. What is the interest of 12416 dollars 25 cents, for 4 months, at the rate of 4 per cent. a year?

89. To what sum will 103 dollars 70 cents amount, in 1 year 2 months and 13 days, at 7 per cent. a year?

90. To what sum will 86 dollars 21 cents amount, in 1 year 1 month and 27 days, at 7 per cent. a year?

91. What is the interest of 502 dollars 9 cents, for 1 year 3 months and 7 days, at 7 per cent. a year?

92. What is the interest of 319 dollars 27 cents, for 2 years 7 months and 11 days, at 7 per cent. a year?

93. What is the amount of 753 dollars 50 cents, for 1 year 9 months and 21 days, at 30 per cent. a year?

94. To what sum will 207 dollars 8 cents amount, in 1 year 4 months and 5 days, at 7 per cent. a year?

95. What is the interest of 99 dollars 10 cents, for 2 years 1 month and 23 days, at 7 per cent. per annum?

To calculate interest on English money, first reduce the shillings, pence and farthings, to the decimal of a pound; the operation will then be as simple as the operation on Federal money.

96. What is the interest of £17 10s. 6d. for 2 years 6 months, at 4 per cent. a year?

£ s. d.   £   £   £

17 10 6 = 17.525. Then, 17.525 × .10 = 1.7525.

£   £ s. d. qr.

1.7625 = 1 15 0 $2\frac{4}{10}$ *Ans.*

97. What is the interest of £42 18s. 9d., for 1 year 7 months and 15 days, at 5 per cent. per annum?

98. What is the interest of £23 8s. 9d., for 6 years, at 7 per cent. a year?

99. To what sum will £140 12s. $3\frac{1}{2}$d. amount, in 1 year 4 months and 12 days, at 6 per cent. a year?

100. To what sum will £463 19s. 6d. amount, in 2 years and 8 months, at 6 per cent. per annum?

101. What is the interest of £104 16s. $10\frac{1}{2}$d., for 11 months and 27 days, at the rate of 7 per cent. a year?

102. What is the interest of £90 5s. 3d., for 1 year 1 month and 9 days, at 7 per cent. per annum?

103. What is the interest of £512 7s. 4d., for 1 year 2 months and 21 days, at 5 per cent. a year?

104. To what sum will £210 10s. 6d. amount, in 1 year 3 months and 18 days, at 7 per cent. a year?

105. What is the interest of £2148 13s. 3d., for 5 months and 17 days, at the rate of $5\frac{1}{2}$ per cent. a year?

106. What is the interest of £750 4s. 6d., for 2 years 3 months and 20 days, at 7 per cent. a year?

107. To what sum will £70 10s. amount, in 3 years 2 months and 10 days, at $7\frac{1}{2}$ per cent. a year?

108. What is the interest of £803 5s. 7d., for 10 months and 14 days, at the rate of 5 per cent. per annum?   9*

109. To what sum will £13 13s. 6d. amount, in 1 year 11 months and 19 days, at 5 per cent. per annum.

PARTIAL PAYMENTS.

In computing interest on notes, bonds, &c. whereon partial payments have been made, it is customary, when settlement is made in a year, or in less than a year from the commencement of interest, to find the amount of the whole principal to the time of settlement, and also the amount of each payment, and deduct the amount of all the payments from the amount of the principal.

The learner may compute the interest on the following notes; considering the rate to be 6 per cent. per annum, when no other rate is stated.

(110.)   Boston, January 14th. 1833.

For value received, I promise Samuel Burbank Jr. to pay him or order the sum of one hundred and forty-one dollars and eight cents, in three months, with interest afterward.   Horace Chase.

On the back of this note were the following endorsements. May 1st. 1833, received seventy-five dollars. September 14th. 1833, received forty-five dollars. The balance of the note was paid January 14th. 1834. How much was the balance?

First payment,	$75.	2nd. payt. $45.	Principal, $141.08
Interest, 8 m. 14 d.,	3.17	Int., 4 m. .90	Int., 9 m. 6.34
Amount,	$78.17	Amount, $45.90	Amount, 147.42
		78.17	124.07
	Amount of payments, $124.07		Balance, $23.35

(111.)   New York, May 25th. 1833.

For value received, I promise Joseph Day to pay him or order the sum of three hundred and one dollars and forty-seven cents, on demand, with interest.

Attest. John Smith.   .   Samuel Frink.

On the back of this note, the following endorsements were made. July 1st. 1833, received sixty-seven dollars and fifty cents. January 4th. 1834, received forty-eight dollars. April 11th. 1834, received thirty-nine dollars. The balance of this note was paid June 21st. 1834. Required the balance.

(112.)   Philadelphia, June 26th. 1833.

For value received, I promise Charles S. Johnson to pay him or order ninety-three dollars and twenty-eight cents, on demand, with interest.   James Orne.

Attest. Levi Dow.

On this note there were two endorsements, viz. Nov. 5th. 1833, received forty-three dollars and seventy-five cents. Feb. 22d. 1834, received thirty-seven dollars. What was due, May 26th. 1834, when the balance was paid.

(113.)   Baltimore, March 4th. 1832.

For value received, I promise Hay & Atkins to pay them or order the sum of four hundred and three dollars and fifty-six cents, in nine months, with interest afterward.   Homer Chase.

The following endorsements were made on the back of this note. Jan. 1st. 1833, received one hundred and eighty-four dollars. August 18th. 1833, received one hundred dollars. This note was taken up Dec. 1st. 1833. What was the balance then due upon it?

(114.)   Hartford, July 11th. 1831.

For value received we promise Joseph Seaver to pay him or order the sum of two hundred and seventeen dollars and fifty cents, in four months, with interest after that time.   Whiting & Davis.

On this note there were three endorsements: viz. Nov. 16th, 1831, received ninety-three dollars. Feb. 12th. 1832, received fifty dollars. August 2d. 1832, received sixty-seven dollars and seventy-five cents. This note was taken up Oct. 4th. 1832. How much was then due upon it?

(115.)   Burlington, October 1st. 1832.

For value received, we promise Hannum, Osgood, & Co. to pay them or order the sum of seven hundred and fourteen dollars, in three months, with interest afterward.   Mason & Gould.

The following payments were endorsed on the note. January 1st. 1833, received three hundred and sixty-four dollars. May 1st. 1833, received one hundred and twenty-five dollars and fifty cents. August 1st. 1833, received eighty-six dollars. Nov. 1st. 1833, received a hundred and ten dollars. The balance due on this note was paid Jan. 1st. 1834. How much was it?

If settlement is not made, till more than a year has elapsed after the commencement of interest, the preceding mode of computing interest, when partial payments have been made, ought not to be adopted; and indeed it is not in strict conformity with law.

The United States Court, and the Courts of the several States, in which decisions have been made and reported, with the exception of Connecticut and Vermont, and a slight variation in New Jersey, have established a general rule for the computation of interest, when partial payments have been made.   This rule is well expressed in the New York Chancery Reports, in a case decided by chancellor KENT, and here given in the Chancellor's own words, as follows.

" The rule for casting interest, when partial payments have been made, is to apply the payment, in the first place, to the discharge of the interest then due. If the payment exceeds the interest, the surplus goes towards discharging the principal, and the subsequent interest is to be computed on the balance of principal remaining due. If the payment be less than the interest, the surplus of interest must not be taken to augment the principal; but interest continues on the former principal until the period when the payments, taken together, exceed the interest due, and then the surplus is to be applied towards discharging the principal; and interest is to be computed on the balance, as aforesaid."

The interest on the following notes, must be computed by the above legal rule.

(116.)   Washington, March 4th. 1832.
For value received, I promise Nehemiah Adams to pay him or order the sum of one thousand two hundred dollars, on demand, with interest.   Charles Train.
   Attest. William Dorr.

The following endorsements were made on this note. June 10th. 1832, received one hundred and sixty-nine dollars and twenty cents. Oct. 22d. 1832, received twenty dollars. March 30th. 1833, received twenty-eight dollars. Nov. 5th. 1833, received six hundred and eighteen dollars and five cents. What was the balance due, on taking up this note, March 5th. 1834?

Principal, -		$ 1200.
Interest from Mar. 4, to June 10, (3 m. 6 d.),	- -	19.20
First Amount,	- -	1219.20
First payment - - - - -,	- -	169.20
Balance, forming a new principal, - -	- -	1050.00
Interest from June 10, to Oct. 22, (4 m. 12 d.),	$23.10	
Second payment, - - - - -	20.	
Leaving interest unpaid, - - - -	3.10	
Interest from Oct. 22, to Mar. 30, (5 m. 8 d.),	27.65	
	30.75	
Third payment, - - - - -	28.00	
Leaving interest unpaid, - - - -	2.75	
Interest from Mar. 30, to Nov. 5, (7 m. 6 d.),	37.80	40.55
Second Amount,	- -	1090.55
Fourth payment, - - - - -	- -	618.05
Balance, forming a new principal, -	- -	472.50
Interest from Nov. 5, to Mar. 5, (4 m.),	- -	9.45
Balance due on taking up the note, -	- -	$ 481.95

(117.)   Richmond, Jan. 5th. 1833.

For value received, I promise Joseph Tufts to pay him or order one hundred and forty-three dollars and fifty cents, on demand, with interest.   John Hanes.

Two payments were endorsed upon this note : viz. April 13th. 1833, received forty-five dollars and eighty-four cents. Dec. 22d. 1833, received fifty-four dollars and fifteen cents. The balance of this note was paid March 28th. 1834. How much was it ?

(118.)   Raleigh, July 1st. 1832.

For value received, I promise Charles Goodrich to pay him or order the sum of six hundred and twenty-five dollars and fifty cents, in three months, with interest afterward.   John Frink.

Three payments were endorsed upon this note: viz. January 1st. 1833, received two hundred dollars. Nov. 1st. 1833, received twenty dollars. Jan. 1st. 1834, received three hundred dollars. The balance was paid May 1st. 1834. How much was it ?

(119.)   Charleston, Dec. 22d 1830.

For value received, I promised George Winship to pay him or order ninety-seven dollars and eighty cents, on demand, with interest,   Thomas White.

The endorsements made on this note were the following. Oct. 12th. 1831, received twelve dollars eighty-five cents. July 20th. 1832, received twelve dollars and seventeen cents. Feb. 26th. 1833, received fourteen dollars and ninety-five cents. August 26th, 1833, received thirty-six dollars and ten cents. Required the balance, which was paid Jan. 31st. 1834.

(120.)   Augusta, January 1st. 1831.

For value received, I promise Israel Capen to pay him or order eighty-four dollars and forty cents, on demand, with interest.   Edward Ruggles.

On the back of this note were the following endorsements. Oct. 9th. 1831, received nineteen dollars and thirty-two cents. July 15th. 1832, received twenty dollars. April 9th. 1833, received twenty-one dollars and eighty-one cents. Oct. 9th. 1833, received twenty-two dollars and fifteen cents. The balance of this note was paid Feb. 19th. 1834. How much was it?

(121)   New Orleans, Feb. 22d. 1830.

For value received, I promise Maynard and Noyes to pay them or order the sum of nine hundred dollars, in three months, with interest till paid.   Isaac Jettison.

Attest. William Proctor.

The following payments were endorsed upon the note. May 22d. 1830, received twenty-five dollars. Sept. 22d. 1830, received fifteen dollars. May 22d. 1831, received thirty-five dollars. May 22d. 1832, received one hundred and forty-five dollars and twelve cents. Dec. 4th. 1832, received one hundred and twenty-five dollars and sixty cents. May 22d. 1833, received two hundred and nineteen dollars and sixty cents. Dec. 31st. 1833, received two hundred and sixty-eight dollars and twenty-five cents. The balance of this note was paid Feb. 24th. 1834. What was the balance?

(122.)   Cincinnati, Dec. 1st. 1830.

For value received, I promise Horatio Davis to pay him or order the sum of one thousand dollars, on demand, with interest till paid.   Edward Lang.

Five partial payments were endorsed on this note: viz. Feb. 1st. 1832, received seventy-five dollars. June 1st. 1832, received twenty dollars. August 1st. 1833, received twenty dollars. October 1st. 1833, received seven hundred and fifty dollars. Feb. 1st. 1834, received one hundred dollars. The balance of this note was paid June 1st. 1834. How much was it?

(123.)   Louisville, April 4th. 1832.

For value received, I promise Samuel H. Wheeler to pay him or order the sum of three hundred and ninety-six dollars, on demand, with interest, at the rate of 7 per cent. a year, till paid.   George Guelph.

Partial payments were made on this note, as follows: Sept. 14th. 1832, received twelve dollars. May 4th. 1833, received eighteen dollars. Oct. 24th 1833, received forty-nine dollars twelve cents. The balance was paid May 30th. 1834. What was the balance?

(124.)   Nashville, Sept. 7th. 1831.

For value received, I promise Darius Pond to pay him or order the sum of four hundred and eighty-six dollars and ninety cents, on demand, with interest at the rate of 7 per cent. a year.   Martin Smith.

The following partial payments were endorsed on this note. March 22d. 1832, received one hundred and twenty-five dollars. Nov. 29th. 1832, received one hundred and fifty dollars. May 13th. 1833, received one hundred and twenty dollars. The balance was paid April 19th. 1834. Required the balance.

(125.)   Albany, August 13th. 1830.

For value received, I promise Theodore Leonard to pay him or order the sum of two hundred and ninety-eight dollars and nineteen cents, on demand, with interest at the rate of 7 per cent. a year.   Stephen Kirkland.

Attest. W. Stevenson.

The following endorsements were made on this note. April 6th. 1831, received fifty-four dollars. Dec. 17th. 1831, received forty-two dollars. June, 21st. 1832 received sixty-one dollars. Feb. 26th. 1833, received thirty-seven dollars and eighty cents. July 8th. 1833, received seventy-five dollars. The balance was paid May 12th. 1834. How much was the balance?

COMPOUND INTEREST.

Compound interest is that which is paid not only for the use of the principal, but also, for the use of the interest after it becomes due.

When the interest is payable annually, find the interest for the first year, and add it to the principal, and this amount is the principal for the second year. Find the interest on this second principal, and add as before; this amount is the principal for the third year: and so on through the whole number of years. When the interest is payable half-yearly, or quarterly, find the interest for half a year, or a quarter of a year, and add it to the principal, and thus proceed through the whole time. Subtract the first principal from the last amount, and the remainder is the compound interest.

126. What is the compound interest of a thousand dollars for 3 years, at 6 per cent. per annum?

$1000.	principal.
60.	interest for the first year.
1060.	amount, principal for the second year.
63.60	interest for the second year.
1123.60	second amount, principal for third year.
67.416	interest for the third year.
1191.016	third amount.
1000.	first principal deducted.
$191.016	Answer.

127. What is the compound interest of 740 dollars for 6 years, at 6 per cent. per annum?

128. What is the compound interest of 500 dollars, for 4 years, at 7 per cent. per annum?

129.. To what sum will 450 dollars amount, in 5 years, at 5 per cent. per annum, compound interest?

130. What is the compound interest of £760 10s. for 4 years, at 4 per cent. per annum?

131. A gave B a note for 300 dollars, with interest at 6 per cent. a year, payable semiannually. How much did it amount to in 2 years, at compound interest?

132. At compound interest, what will 600 dollars amount to in 1½ year, at the rate of 6 per cent. a year, interest payable quarterly?

PROBLEMS IN INTEREST.

In reviewing the subject of simple interest, we perceive four several problems, which arise from its conditions, and which we shall now distinctly notice.

PROBLEM I. The principal, time, and rate per cent. given, to find the interest.

RULE. *Multiply together the decimal expressing the rate per annum, the time in years and the decimal of a year, and the principal: the product will be the interest.*

This problem has already been exemplified in the preceding pages of this article.

PROBLEM II. The principal, time, and amount given, to find the rate per cent. per annum.

RULE. *Subtract the principal from the amount, and the remainder will be the interest for the given time. Divide this interest by the given time expressed in years or the decimal of a year, and the quotient will be the interest for one year. Divide the interest for one year by the given principal, and the quotient will be the rate per cent. per annum.*

133. At what rate per cent. per annum must 172 dollars 40 cents be put on interest, in order to amount to 332 dollars 74 cents, in 5 years?

134. Lent 51 dollars 25 cents, and in 1 year and 4 months it amounted to 55 dollars 35 cents. What was the rate per cent. per annum?

135. Borrowed 340 dollars for 9 months, and at the expiration of the time it amounted to 355 dollars 30 cents. What was the rate of interest per annum?

136. At what rate per cent. per annum must 87½ cents be put on interest, in order to amount to 98 cents, in 2 years?

PROBLEM III. The principal, rate per cent., and amount given, to find the time.

RULE. *Subtract the principal from the amount, and the remainder will be the interest. Divide the interest by the principal, and the quotient will be the interest of 1 dollar. Divide the interest of 1 dollar by the rate, and the quotient will be the time.*

137. In what time will 89 dollars 25 cents amount to 92 dollars 82 cents, at the rate of 6 per cent. a year?

138. In what time will 171 dollars 40 cents, amount to 231 dollars 39 cents, at 7 per cent. a year?

139. Borrowed 163 dollars 50 cents at 6 per cent. a year; at the time of payment it amounted to 176 dollars 58 cents. How long did I keep the money?

140. In what time will 4810 dollars 25 cents, amount to 5002 dollars 66 cents, at 6 per cent. a year?

141. Lent 114 dollars at an interest of 7 per cent. a year; on its return it amounted to 127 dollars 30 cents. How long was it out?

142. In what time will $100, or any other sum of money double, at the rate of 6 per cent. per annum, simple interest?

PROBLEM IV. The amount, time, and rate per cent. given, to find the principal.

RULE. *Divide the amount by the amount of 1 dollar for the time, and the quotient will be the principal.*

This problem forms the subject of the next article, under the head of *Discount.*

XVI.

DISCOUNT.

DISCOUNT is an allowance made for the payment of money before it is due.

The *present worth* of a debt, payable at a future period without interest, is that sum of money, which, being put on interest, would amount to the debt, at the period when the debt is payable.

• It is obvious, that, when money is worth 6 per cent. per annum, the present worth of $1.06, payable in a year, is $1. Hence, the present worth of any debt, payable in a year, is as many dollars as there are times $1.06 in the debt. And hence we deduce the following.

RULE. *Divide the debt by the amount of 1 dollar for the time, and the quotient is the present worth. Subtract the present worth from the debt, and the remainder will be the discount.*

1. What is the present worth of 450 dollars, payable in 6 months, when money is worth 6 per cent. per annum?

2. What is the present worth of 535 dollars, payable in 15 months, when money is worth 6 per cent. per annum?

3. When money is let for 6 per cent. per annum, what is the present worth of a note for 1530 dollars, payable in 18 months?

4. Sold goods to the amount of 1500 dollars, to be paid one half in 9 months, and the other half in 18 months: what is the present worth of the goods, allowing interest to be 5 per cent. per annum?

5. What is the present value of a note for 2576 dollars and 83 cents, payable in 9 months, when interest is 6 per cent. per annum?

6. When interest is 6 per cent. a year, what is the difference between the discount on 1285 dollars for a year and 8 months, and the interest of the same sum for the same time?

7. Purchased goods amounting to 6568 dollars 50 cents on a credit of 8 months: allowing money to be worth 4 per cent. a year, how much cash down will pay the bill?

8. A man, having a horse for sale, was offered for it 225 dollars, cash in hand, or 230 dollars payable in 9 months: he chose the latter, although money was worth 7 per cent. a year. How much did he lose by his ignorance?

9. Bought a quantity of goods for 1831 dollars 53 cents cash, and the same day sold them for 1985 dollars 48 cents on a credit of 6 months, when money was 5 per cent. a year. How much did I gain upon the goods?

10. What is the discount on 198 dollars 60 cents, for 9 months, when interest is 5 per cent. a year?

11. What is the discount on 241 dollars 81 cents, for 7 months, when interest is $4\frac{1}{2}$ per cent. a year?

12. What is the present worth of 741 dollars 65 cents, payable in 48 days; interest being 6 per cent.?

XVII.

BANKING.

A BANK is an institution which trafficks in money. It is owned in shares, by a company of individuals, called *stockholders*; and its operations are conducted by a President and board of Directors. It has a deposite of specie, and issues *notes* or *bills*, which are used for a circulating medium, as money. These bills are mostly obtained from the bank in loans, on which interest is paid; and the amount of bills issued being greater than the amount of specie kept in deposite, a profit accrues to the bank.

The interest on money hired from a bank, is paid at the time when the money is taken out—the hirer receiving as much less than the sum he promises to pay, as would be equal to the interest of what he promises to pay, from the time of hiring the money until the time it is to be paid. From this circumstance, the interest on money hired from a bank is called *discount*, and the promissory note received at the bank is said to be *discounted*.

A note, to be discounted at a bank, is usually made payable to some person, who endorses it, and who thereby binds himself to pay the debt, in case the signer of the note should fail to do so. Any person, therefore, who holds the note of another, payable at a future time, may endorse it, and obtain the money for it at a bank, by paying the bank discount; provided the credit of the parties is undoubted.

It is customary in banks, to compute the discount on every note for 3 days more than the time stated in the note; and the debtor is not required to make payment until 3 days after the stated term of time has elapsed. These 3 days are called *days of grace*.

1. What is the bank discount on 775 dollars for 30 days, and grace, when interest is 6 per cent. a year?

2. What is the bank discount on 900 dollars for 90 days, and grace, at the rate of 6 per cent. a year?

3. How much is received on a note for 2540 dollars 80 cents, payable in 4 months, discounted at a bank, when interest is 4½ per cent. a year ?

4. A note for 452 dollars, payable in 7 months, is discounted at a bank, when interest is 6 per cent. per annum.   What sum is received on it ?

5. A note for 3000 dollars, payable in 70 days, is discounted at a bank, when interest is 6 per cent. a year. What sum is received on it ?

6. A merchant bought 1625 barrels of flour for 5 dollars a barrel cash, and on the same day sold it for 5 dollars 50 cents a barrel, on a credit of 8 months, took a note for the amount, and got it discounted at a bank, when money was 6 per cent. a year.   How much did he gain on the flour ?

7. A man got his note for $1000, payable in 3 months, discounted at a bank, at the rate of 6 per cent., and immediately put the money he received for his note on interest for 1 year, at 6 per cent.   He kept the money from the bank 1 year, by renewing his note every 3 months, and paying in the required bank discount at each renewal.   At the end of the year he received the amount of the money he had put on interest, and paid his note at the bank.   How much did he lose by this exchange ?

In the above example, interest on the several discounts paid into the bank forms part of the loss.

8. A money broker subscribed for 20 shares in a new bank; at $100 a share.   When the bank commenced operation he paid in 50 per cent. of the price of his stock, and in 6 months after, he paid in the remainder.   In 12 months from the time the bank commenced, there was a dividend of 3½ per cent. on the stock among the stockholders; and the same dividend accrued every 6 months thereafter.   At the end of 3 years the broker sold his stock at 7 per cent. advance.   Now, allowing that this broker hired his money, and paid 6 per cent. annually, how much did he make by the speculation ?

In this example, the broker must charge annual interest on the interest he pays, and must give credit for annual interest on his share of the dividends.

10*

XVIII.

EQUATION OF PAYMENTS.

EQUATION OF PAYMENTS consists in finding a mean time for the payment at once of several debts, payable at different times, so that no loss of interest shall be sustained by either party.

For instance, if A owes B one dollar, payable in 2 months, another dollar payable in 3 months, and a third dollar payable in 4 months, at what time may the three sums be paid at once, without injustice to either of them ? It is evident, that the interest of 1 dollar for 2 months, is the same as the interest of 2 dollars for 1 month; and the interest of 1 dollar for 3 months, is the same as the interest of 3 dollars for 1 month; and the interest of 1 dollar for 4 months, is the same as the interest of 4 dollars for 1 month: 2 dollars, 3 dollars, and 4 dollars, added together, make 9 dollars for 1 month; but the three sums to be paid, when added together, make only 3 dollars, which sum being only a third part of 9 dollars, the term of credit must be three times as long, or 3 months, which is the equated time. This result is obtained by multiplying the sum, payable in 2 months, by 2; that payable in 3 months, by 3; and that payable in 4 months, by 4; and then adding the several products together, and dividing the sum of them by the sum of the debts.

RULE. *Multiply each debt by the time, in which it is payable, and divide the sum of the products by the sum of the debts: the quotient will be the equated time.*

1. If I owe you 50 dollars payable in 4 months, 75 dollars payable in 6 months, and 100 dollars payable in 7 months, in what time may the three sums be paid at once, without loss to either of us ?

2. A owes B 200 dollars, 40 dollars of which is to be paid in 3 months, 60 dollars in 5 months, and the remainder in 10 months. At what time may the whole be paid at once, without injustice to either party ?

3. Bought goods to the amount of 1552 dollars, payable at four different times, as follows; 225 dollars and 75 cents in 4 months, 250 dollars and 25 cents in 6 months, 425 dollars and 50 cents in 8 months, 650 dollars 50 cents in 10 months; but afterward agreed with my creditor to pay him all at once, at the equated time. What was the time?

4. If I owe you three sums of money payable at different times, viz. 50 pounds in six months, 60 pounds in 7 months, and 80 pounds in 10 months, what is the equated time for paying the whole at once?

5. Bought goods to the amount of 1000 dollars, 200 dollars of which was to be paid down, 400 dollars in 5 months, and the remainder in 15 months; but it was afterward agreed, that the whole be paid at once. In what time ought the payment to be made?

6. A merchant has due to him a certain sum of money, to be paid as follows; $\frac{1}{4}$ in 2 months, $\frac{1}{3}$ in 3 months, and the rest in 6 months. What is the equated time for paying the whole?

7. Sold goods amounting to 1296 dollars, of which 346 dollars was to be paid in 2$\frac{1}{2}$ months, 323 dollars in 6 months, and the balance in 10 months; but the purchaser afterward agreed to make but one payment of the whole. What term of credit ought he to have?

8. Bought goods to the amount of 640 dollars 80 cents, payable $\frac{1}{4}$ down, $\frac{1}{4}$ in 4 months, $\frac{1}{4}$ in 8 months, and the balance in a year; but afterward made an agreement to pay the whole at one time. In what time ought I to pay for the goods?

9. A merchant has due to him $300 to be paid in 60 days, $500 to be paid in 120 days, and $750 to be paid in 120 days. What is the equated time for these dues?

10. A owes B $1200, to be paid in 8 months; but A offers to pay $400 in 4 months, on condition that the remainder shall continue unpaid an adequate term of time. In what time ought the remainder to be paid?

11. If a debt of $1000 be payable at the end of 7 months, and the debtor agree to pay $300 at present, what is the proper time for paying the rest?

XIX.

PROFIT AND LOSS.

The ascertaining what is gained or lost in buying and selling, and the adjusting of the price of goods so as to gain or lose a certain sum, or a certain per cent., come under the head of *Profit and Loss.*

1. Bought a piece of broadcloth containing 28 yards for 112 dollars, and sold it at 5 dollars 25 cents a yard. How much, and what per cent. was my profit? (See ART. XIV, Example 45.)

2. Bought 3 pieces of broadcloth, containing 28 yards each, at 5 dollars 25 cents a yard. At what price per yard must I sell it, to gain 20 per cent.?

3. Bought cloth at 4 dollars 60 cents a yard, which, not proving so good as I expected, I sold at 3 dollars 91 cents a yard. What per cent. did I lose?

4. Bought 1250 barrels of flour for 6250 dollars. At what price per barrel must I sell it, to make a profit of $12\frac{1}{2}$ per cent.?

5. Bought 30 hogsheads of molasses, at 20 dollars a hogshead, in Havana; paid duties 20 dollars 66 cents; freight 40 dollars 78 cents; porterage 6 dollars 5 cents; insurance 30 dollars 84 cents. What per cent. shall I gain by selling at 26 dollars per hogshead?

6. Bought wheat at 75 cents a bushel; at what price per bushel must I sell it, to gain 20 per cent.?

7. A merchant received from Lisbon 180 casks of raisins, containing $80\frac{3}{5}$ lb. each, which cost him 2 dollars 18 cents a cask. At what price per cwt. must he sell them, to gain 25 per cent.?

8. If I sell sugar at 8 dollars per cwt., and thereby lose 12 per cent., what per cent. do I gain or lose, by selling the same at 9 dollars per cwt.?

9. If I purchase 6 pipes of wine for 816 dollars, and sell it at 59 dollars 50 cents a hhd. do I gain or lose, and what per cent.?

10. If you purchase 5 cwt. 1 qr. 12 lb. of rice, at 2 dollars 80 cents per cwt., at what price per pound must you sell it, to make 6 dollars on the whole?

11. If I purchase 13 cwt. of coffee at $12\frac{1}{2}$ cents per pound, at what price per lb. must I sell it, to gain 80 dollars 8 cents on the whole?

12. A miller sold a quantity of corn at 1 dollar a bushel, and gained 20 per cent.; soon after, he sold of the same, to the amount of $37.50, and gained 50 per cent. How many bushels were there in the last parcel, and at what did he sell it per bushel?

XX.

PARTNERSHIP.

PARTNERSHIP is the union of two or more individuals in trade. The company thus associated is called a firm: and the amount of property, which each partner puts into the firm, is called his stock in trade. The profit or loss is shared among the partners, when the stock of each is employed an equal length of time, in proportion to each partner's stock in trade; but, when the stock of the several partners is employed in the firm unequal terms of time, in proportion to each one's stock and the time it was employed.

1. A, B, and C entered into partnership, and the stock of each was employed in the firm one year. A put in 240 dollars, B 360 dollars, and C 120 dollars. They gained 350 dollars. What was each partner's share of the gain?

We find in this example, that the whole capital of the firm was 720 dollars. A's stock was 240 dollars, and he must have $\frac{240}{720}$ of the gain. B's stock was 360 dollars, and he must have $\frac{360}{720}$ of the gain. C's stock was 120 dollars, and he must have $\frac{120}{720}$ of the gain. Observe the following statement.

$\frac{240}{720} = \frac{2}{6}$; and $\frac{2}{6}$ of $350. is $116\frac{2}{3}$ dollars, A's share.

$\frac{360}{720} = \frac{3}{6}$; and $\frac{3}{6}$ of $350. is 175 dollars, B's share.

$\frac{120}{720} = \frac{1}{6}$; and $\frac{1}{6}$ of $350. is $58\frac{1}{3}$ dollars, C's share.

$350   Proof.

2. Messrs. Ralph Wheeler, Samuel Slade, and James Libbey formed a connexion in business under the firm of Wheeler, Slade, and Libbey. Wheeler put into the firm 2500 dollars; Slade 2000 dollars; and Libbey 1500 dollars. The stock of the several partners was in trade the same term of time, and they gained 1500 dollars. What was each partner's share of the profit?

3. Messrs. Joel Haven, Israel Varnum, Tyler Penniman, and James Conant formed a partnership under the firm of Haven, Varnum, and Co. Haven put into the firm 4000 dollars, Varnum 2500 dollars, Penniman 1500 dollars, and Conant 750 dollars. They traded in partnership 3 years, and gained 1750 dollars. How much was each partner's share of gain?

4. A, B, C, and D traded together one year. A put in 800 dollars, B 500 dollars, C 300 dollars, and D 150 dollars; but by misfortune they lost 350 dollars. What loss did each partner sustain?

5. A gentleman dying, left two sons and a daughter, to whom he bequeathed the following sums; viz. to the elder son 1200 dollars, to the younger, 1000 dollars, and to the daughter 800 dollars; but it was found that his whole estate amounted only to 750 dollars. How much did each child receive from the estate?

6. Three merchants bought a ship, for which they gave 8000 dollars. A paid 2850 dollars, B 1980 dollars, and C the rest: in her first voyage she cleared 6400 dollars. How much of the profit had each partner?

7. A and B traded together. A put into the firm 540 dollars, and B the rest: they gained 387 dollars, of which B's share was 225 dollars. What was A's gain, and what was B's stock?

8. The capital stock in the firm of Farmer, Turner and Hancock, was 18477 dollars 60 cents. Farmer's stock was 9238 dollars 80 cents; Turner's 6929 dollars 10 cents; and Hancock's the remainder. The stock

of the several partners was in the firm the same term of time: by misfortunes of various kinds, they lost 12375 dollars 20 cents. What loss did each partner sustain?

9. A, B, and C traded in partnership. A's stock was 385 dollars 50 cents; B's 297 dollars 75 cents; C's 175 dollars 25 cents: they gained 343 dollars 40 cents. What was each one's share of gain?

When the stock of the several partners is in the firm unequal terms of time, the profit or loss must be apportioned with reference both to stock and time. Thus, A, B, and C, traded in company; A put in 200 dollars for 3 months, B 180 dollars for 5 months, and C 70 dollars for 10 months: they gained 132 dollars. Now, to apportion this gain justly, we say that A's 200 dollars for 3 months was the same as 600 dollars for 1 month; B's 180 dollars for 5 months, the same as 900 dollars for 1 month; and C's 70 dollars for 10 months the same as 700 dollars for 1 month; therefore it is the same as if A had put in 600 dollars, B 900 dollars, and C 700 dollars, all for an equal term of time. These sums added together make 2200 dollars; therefore, A had $\frac{600}{2200}$ of the gain, B $\frac{900}{2200}$, and C $\frac{700}{2200}$. These fractions, when reduced, are $\frac{6}{22}$, $\frac{9}{22}$, and $\frac{7}{22}$. $\frac{1}{22}$ of 132 dollars is 6 dollars; then A had 6 times 6 dollars, B 9 times 6 dollars, and C 7 times 6 dollars.

RULE. *Multiply each partner's stock by the time it was in the firm; make each product the numerator of a fraction, and the sum of the products a common denominator; then multiply the whole gain or loss by each of these fractions, for each partner's share.*

10. A, B, and C traded in company. A put in 400 dollars for 9 months, B 300 dollars for 6 months, and C 200 dollars for 5 months: they gained 320 dollars. What was the gain of each?

11. X, Y, and Z formed a partnership. X put into the firm 500 dollars for 18 months, Y 380 dollars for 13 months, and Z 270 dollars for 9 months; but they lost 818 dollars 50 cents. What was the loss of each?

12. R and S entered into partnership for 16 months; R put in at first $1200, and at the end of 9 months $200 more. S put in at first $1500, and after 6 months had elapsed he took out $500. In this partnership they gained $772.20. How must the gain be divided?

13. On the first day of January, A began business with 380 dollars; on the first day of May following, he took B into partnership with 270 dollars; on the first day of the next August, they took in C with 400 dollars: at the end of the year, they found there was a gain of 436 dollars. What share of the gain had each?

14. Gould and Davis entered into partnership for one year. Gould's stock, at first, was only 500 dollars, but at the end of 5 months he put in 150 dollars more. Davis's stock, at first, was 600 dollars, but at the end of 9 months he took out 200 dollars: at the end of the year, it was found they had gained 682 dollars 50 cents. What was the gain of each partner?

15. Three farmers hired a pasture at 60 dollars 50 cents for the season. A put in 5 cows $4\frac{1}{2}$ months, B 8 cows 5 months, and C 9 cows $6\frac{1}{2}$ months. What rent did each pay?

16. A and B hired a coach in the city, to go 40 miles for $20, with liberty to take in two more passengers. When they had ridden 15 miles they admitted C; and on their return, within 25 miles of the city, they admitted D. As each person is to pay in proportion to the distance he rode, it is now required to settle the coach hire justly between them.

17. Messrs. Howard, Bender, Dorr, and Tremere were partners for 2 years, under the firm of Joseph Howard and Co. When the firm commenced business, Howard's stock was 6000 dollars, Bender's 3500 dollars, Dorr's 2800 dollars, and Tremere's 1700 dollars. At the end of 8 months, Howard withdrew 3000 dollars from the firm; after trading 10 months, Tremere added 1300 dollars to his former stock; at the end of the first year, Bender withdrew 800 dollars. At the close of the two years, they had gained 3608 dollars 40 cents. How much was each partner's share of the gain?

XXI.

BANKRUPTCY.

In the course of mercantile business, it happens not unfrequently, that a merchant, either from misfortune or imprudence, becomes insolvent, and what property he has, is distributed among his creditors in proportion to their respective dues.

Questions in bankruptcy are performed on the same principle with those in partnership: we first ascertain the amount of the bankrupt's property, then the amount of his debts, next what per cent. he pays, and then multiply the sum due to each creditor by the decimal expressing the per cent. which the debtor pays.

1. A man failing, owed the following sums: to A 120 dollars 68 cents, to B 150 dollars 75 cents, to C 310 dollars 32 cents, to D 208 dollars 25 cents; and his whole property amounted to only 632 dollars, which was divided among them in proportion to their respective demands. How much did each receive?

2. If the money and effects of a bankrupt amount to 3361 dollars 74 cents, and he is indebted to A in the sum of 1782 dollars 24 cents, to B 1540 dollars 76 cents, and to C 2371 dollars 17 cents, how much will each of them receive?

3. A person failing in trade, owed A 539 dollars, B 756 dollars 80 cents, C 854 dollars 16 cents, and D 1200 dollars; his property amounted to 837 dollars 49 cents, which was distributed among them in proportion to their several demands. How much did each creditor lose by the failure?

4. A bankrupt owes A 813 dollars 74 cents, B 3673 dollars 46 cents, C 1840 dollars 40 cents, D 117 dollars 80 cents, and E 814 dollars 60 cents, his whole property, worth 4029 dollars 30 cents, he gives up to his creditors. What per cent. does he pay, and how much does each creditor receive?

11

XXII.

ASSESSMENT OF TAXES.

TAXES are imposts paid by the people for the support of government. They are assessed on the citizens in proportion to their property; except the poll tax, which is assessed by the head without regard to property.

An accurate inventory of all the taxable property of every citizen is indispensable to a just assessment of taxes, and is the first thing to be obtained.

When a tax is to be assessed on property and polls, we first ascertain the amount which the polls pay, and deduct it from the sum to be raised; then apportion the remainder according to each man's property.

To effect the apportionment, we find what per cent. of the whole property to be taxed, the sum to be raised is; then multiply each man's inventory by that per cent. expressed in decimals, and the product is his tax.

1. A tax of four hundred and fifty dollars is to be assessed on a parish, in which there are 40 rateable polls. Of this tax, that to be assessed on the polls amounts to 50 dollars; leaving 400 dollars to be assessed on the real and personal property of the parish, which by inventory is 40000 dollars. What must a parishioner pay, whose real estate in the inventory is 700 dollars, personal property 150 dollars, and who pays for one poll?

400 dollars is one per cent. (.01) of 40000 dollars.

Real estate	700
Personal estate	150
Total property	850
	.01
Tax on property	8.50
Poll tax	1.25
Total tax	$9.75 *Ans.*

2. An incorporated town, in which the real and personal property was valued in the inventory at 72856 dollars, and

in which there were 154 rateable polls, voted to raise 1285 dollars 34 cents by taxation. The tax on each poll was 1 dollar 25 cents. How much did A pay, whose real estate stood in the inventory at 2146 dollars, personal property at 224 dollars, and who paid for 1 poll? How much did B pay, whose real estate in the inventory was 1000 dollars, his personal property 140 dollars, and who paid for 2 polls? How much did C pay, whose real estate was valued at 785 dollars, personal property at 103 dollars, and who paid for 3 polls? How much did D pay, whose real estate in the inventory was 4000 dollars, personal property 478 dollars, and who paid for 1 poll? How much did E pay, who had no real estate, whose personal property was valued at 250 dollars, and who paid for 4 polls? How much did a single woman pay, whose real estate was valued at 500 dollars, and her personal property at 120 dollars?

Assessors find it most expedient to make a table, which shall exhibit at once, the tax on all sums, from $1 up to any amount required. The table is made by multiplying the per cent. which the tax amounts to, by the several numbers, 1, 2, 3, 4, and so on.

The following is a table of taxes to be made when $1\frac{1}{2}$ per cent. is to be raised on the valuation of property.

$1	pays	.015	$20	pay	.30	$200	pay	$3.00
2	"	.03	30	"	.45	300	"	4.50
3	"	.045	40	"	.60	400	"	6.00
4	"	.06	50	"	.75	500	"	7.50
5	"	.075	60	"	.90	600	"	9.00
6	"	.09	70	"	1.05	700	"	10.50
7	"	.105	80	"	1.20	800	"	12.00
8	"	.12	90	"	1.35	900	"	13.50
9	"	.135	100	"	1.50	1000	"	15.00
10	"	.15						

3. By the above table, what would be the tax on $6425 real estate, and $2346 personal estate?

4. By the above table, what would be the tax of a freeholder, whose real estate is valued at $9842, and personal estate, at $15066; poll tax $1.25?

XXIII.

GENERAL AVERAGE.

Whenever a ship is in distress, and the master deliberately makes a sacrifice of any part of the lading, or of the ship's furniture, masts, spars, rigging, &c. for the preservation of the rest, all the property on board, which is saved by the sacrifice, must contribute towards the value of what is thus sacrificed. The contribution is called a GENERAL AVERAGE ; and the property sacrificed called the *jettison*.

If a vessel is accidentally stranded, and by extraordinary labor and expense is set afloat, and enabled to complete her voyage with the cargo on board, the expense bestowed for this object also becomes a subject for general average.

When a vessel has been forced by accidents, arising from the perils of the sea, to enter a port in order to repair, all the charges incurred in consequence, together with the wages and provisions of the master and crew during the delay, are brought into a general average.

The contributory interests are the ship, the cargo, and the freight; and these must be cleared of all charges attached to them, before the average is made.

The contributory value of freight, in the ports generally, is ascertained by deducting one third of the gross freight; in New York, however, one half is deducted. This deduction is made for seamen's wages.

In computing a general average for masts, rigging, &c. a deduction of one third is made from the cost of replacing them; because the new articles are supposed to be so much better than the old ones.

Particular average is nothing more than a partial loss, and is borne wholly by the owner of the property damaged. In making a general average, the partial loss, or particular average, is deducted from the original value of the damaged property, and the remainder contributes to the general average.

The jettison contributes to a general average; otherwise its owner would not share in the general loss.

1. The brig Ceres sailed from Gottenburg on the 24th. of August 1833, bound for Boston with a cargo of iron and steel. She suffered considerable damage from tempestuous weather, and arrived in Boston harbor October 13th., where she got aground on Williams's Island, and was obliged to discharge part of her cargo in lighters, in order to get the vessel off. After this she was moored in safety at a wharf.

The expense of lightening the vessel, to get her afloat, was 106 dollars and one cent, and was borne by a general average.

From the surveyor's report, it appears that the damage sustained by the vessel on the voyage was 1195 dollars 73 cents. The premium for insurance was 304 dollars 27 cents.

Contributory interests.

Vessel, valued at	$ 8000.	
Less, damage and premium,	1500.	6500.
Freight, gross amount,	1276.96	
Less ⅓, as usual, for seamen's wages,	425.65	851.31

Cargo, shipped by

James Fullerton & Co.	$273.82	
Joshua Crane	626.67	
John Bradford	10378.48	
Wm. Parsons	1144.32	12423.29

$\frac{10601}{1977460} = .00536 +$ $19774.60

Apportionment of contribution.

Vessel, (6500 by .00536) pays	$34.84	
Freight, 851.31 " " "	4.56	
Cargo, 12423.29 " " "	66.61	$106.01

James Fullerton & Co.	$1.46
Joshua Crane	3.37
John Bradford	55.64
Wm. Parsons	6.14
	$66.61

11*

2. The schooner Crescent, on her passage from East-
port to New York, sustained so much damage, that she
was obliged to put into Plymouth to repair. The expenses incurred by putting into this harbor, viz. the pilot-
age, protest, dockage, commission, wages and provision
of the master and crew while in harbor, amounting to
$73.18, were paid by a general average, made in New
York, on the arrival of the vessel in that port.

The vessel was valued at $4500, and the premium and
repairs were estimated at $900. The gross freight was
$153.

Cargo, shipped by E. Foster $ 600.
     Greason and Haughton   240.
     Gold and Tucker   210.
     Bucknam and Gunnison   400.
     Samuel Whaler   160.
     Buck and Hammond   221.37

What per cent. of the contributory interests was the
general average? How much did each of the interests,
and each of the shippers pay?

3. Ship Coral, on her passage from Boston to New
Orleans, grounded at the bar of the river Mississippi,
threw overboard part of her cargo to lighten, when near
the breakers; broke an anchor, anchor stock, and windlass by strain in heaving off, and took a steamboat to
bring her into port, while in this disabled situation.

Statement of loss to be made up by a general average.

A. Howard's goods, thrown overboard,	$925.00
Expense of steamboat,	100.00
Freight lost in consequence of jettison,	17.78
Damage to cable in heaving off,	50.00
Anchor broke and lost,	150.00
All other damage,	57.00
Protest $14. Adjusting average $50.	64.00
	$1363.78
Agent's commission 5 per cent.	68.19
Amount to be made up by the average,	$1431.97

Contributory interests.

Ship valued at N. Orleans, in cash,   . $11000.
Freight, gross amount $862.48, less ⅓,   574.99
Cargo shipped by Bridge & More $18135.

How & Mears	17000.
Gray & Bellows	14680.
James Russell	3670.
A. Howard	925.
$54410.	54410.

Amount of contributory interests, $65984.99

What is the loss per cent.? How much does the ship, how much does the freight, and how much does each of the shippers contribute to the general average?

XXIV.

CUSTOM-HOUSE BUSINESS.

In every port from which merchandise is exported to foreign countries, and into which foreign merchandise is imported, there is an establishment under the direction of the government, called a *Custom House.* The object of this establishment is, to execute the laws of the United States in the collection of *duties* imposed on certain articles of imported merchandise, and on the tonnage of vessels employed in commerce.

In order to secure the exact collection of duties, the law provides, that the cargoes of vessels employed in foreign commerce, shall be inspected, and weighed or gauged by the custom-house officers. In the custom-house weight and gauge of goods, certain allowances are made on account of the box, cask, bag, &c. containing the goods, and on account of leakage, breakage, &c.

ALLOWANCES.

Draft is an allowance made from the weight of each box, bag, cask, &c. of goods, on account of probable waste.

Tare is an allowance made for the weight of the box, bag, cask, &c. containing the goods.

The whole weight of any parcel of goods, including the weight of the box, bag, cask, &c. containing the goods, is called the *gross weight*.

The weight of any parcel of goods after the draft and tare have been deducted, is called the *neat weight*.

The allowance for draft is stated in the following table.

On a single box, &c. weighing 1 cwt. or 112 lb.   1 lb.
"   weighing above 1 cwt. and under 2 cwt.   2 lb.
"   "   weighing 2 cwt. and under 3 cwt.   3 lb.
"   "   "   3 cwt. and under 10 cwt.   4 lb.
"   "   "   10 cwt. and under 18 cwt.   7 lb.
"   "   "   18 cwt. and upwards,   9 lb.

Observe, that the tare is computed on the remainder of any weight, after the draft has been allowed: and in casting tare, any remainder, which does not exceed half a pound, is not reckoned; but, if it exceed half a pound, it is reckoned a pound.

The tare on sugar in casks, (except loaf) is 12 per cent.
"   on sugar in boxes,   -   -   -   15 per cent.
"   on sugar in bags or mats,   -   -   5 per cent.
"   on cheese in hampers or baskets,   10 per cent.
"   on cheese in boxes,   -   -   20 per cent.
"   on candles in boxes,   -   -   8 per cent.
"   on chocolate in boxes,   -   -   10 per cent.
"   on cotton in bales,   -   -   -   2 per cent.
"   on cotton in ceroons,   -   -   6 per cent.
"   on glauber salts in casks,   -   -   8 per cent.
"   on nails in casks,   -   -   -   3 per cent.
"   on pepper in casks,   -   -   -   12 per cent.
"   on pepper in bales,   -   -   -   5 per cent.
"   on pepper in bags,   -   -   -   2 per cent.
"   on sugar candy in boxes,   -   -   10 per cent.
"   on soap in boxes,   -   -   -   10 per cent.
"   on shot in casks,   -   -   -   3 per cent.
"   on twine in casks,   -   -   -   12 per cent.
"   on twine in bales,   -   -   -   3 per cent.

Tare on all other goods paying a specific duty, is allowed according to the statement of the same in the invoice of the goods, which is considered the actual weight of the box, bag, cask, &c.

The importer may always have the invoice tare allowed, if he make his election at the time of making his entry, and obtain the consent of the collector and naval officer.

For leakage, 2 per cent. is allowed on the gauge, on all merchandise in casks paying duty by the gallon.

For breakage, 10 per cent. is allowed on all beer, ale, and porter in bottles, and 5 per cent. on all other liquors in bottles; or the importer may have the duties computed on the actual quantity by tale, if he so chooses at the time of entry.

The common-size bottles are estimated at the Custom-house to contain $2\frac{2}{3}$ gallons per dozen.

1. What is the neat weight of 40 hogsheads of sugar weighing gross 8 cwt. 3 qr. each; draft and tare as in the tables ?

$$40 \times 4 = \quad \begin{array}{r} 39200 \text{ lb. gross weight.} \\ 160 \text{ lb. draft.} \\ \hline 39040 \end{array}$$

12 per cent. of 39040 is 4684.8 $\begin{array}{r} 4685 \text{ lb. tare.} \\ \hline 34355 \text{ lb. neat weight.} \end{array}$

2. What is the neat weight of 25 bags of pepper, weighing gross 1 cwt. each; draft and tare as in the tables ?

3. Find the neat weight of 6 chests of Souchong tea, weighing gross 98 lb. each, tare 22 lb. per chest.

4. Find the neat weight of 12 casks of raisins, weighing gross 130 lb. each; draft as in the table, tare 12 lb. per cask.

5. What is the neat weight of 8 chests of green tea; gross weight 102 lb. each, tare 20 lb. per chest ?

6. What is the neat weight of 9 bags of coffee, weighing gross 114 lb. each, draft as in table, tare 2 per cent. ?

7. What is the neat weight of 4 casks of glauber salts, gross weight as follows; the first 150 lb.; the 2d. 175 lb.; 3d. 228 lb.; 4th. 264 lb.; draft and tare as in tables ?

8. What is the neat weight of 4 hogsheads of madder; weighing gross 11 cwt. 2 qr. each; draft being allowed as in the table, tare 1 cwt. 2 qr. per cask?

DUTIES.

The duties paid on goods imported from foreign countries into the United States, are either *ad valorem* or specific.

The *ad valorem* duty is a certain per cent. of the actual cost of the goods in the country from which they are brought.

The specific duty is fixed at a certain sum per ton, hundred weight, pound, gallon, square yard, &c.

Observe that the allowances for tare, draft, &c. are to be made, before the duties are computed.

9. What is the duty on an invoice of silk goods, which cost in Canton 4836 dollars, at 10 per cent. ad valorem?

10. What is the duty on an invoice of woollen goods, which cost in England 5729 dollars, at 44 per cent. ad valorem?

11. Compute the duty on 6 boxes of chocolate, weighing gross 1 cwt. per box; draft and tare as in the tables; duty 4 cents per lb.

12. Cast the duty on 12 boxes of Windsor soap; gross weight 84 lb. per box; cost in England 1 dollar per lb.; tare as in the table; duty 15 per cent.

13. Calculate the duty on 5 boxes brown Havana sugar; gross weight as follows; the first, 7 cwt. 2 qr.; 2d. 8 cwt. 3 qr.; 3d. 9 cwt. 1 qr.; 4th. 10 cwt. 3 qr. 20 lb.; 5th. 11 cwt. 1 qr. 14 lb.; draft and tare as in the tables; duty 2½ cents per lb.

14. What is the duty on a cargo of 148 tons of iron, at 30 dollars per ton?

15. Compute the duty on 4 pipes of wine; allowance for leakage as in the table; duty 7½ cents per gallon.

16. Cast the duty on 10 gross of London porter; allowance for breakage as in the table; duty 20 cents per gallon.

17. What is the duty on 10 boxes of Spanish cigars, containing 1100 each; duty $2.50 per 1000?

18. Compute the duty on 4 casks of Rochelle salts, invoiced at $10 per cwt.; gross weight of 1st cask 1 cwt. 2qr. 12lb.; 2d. 1cwt. 1 qr. 17lb.; 3d. 2cwt. 3qr. 7lb.; 4th. 4cwt. 1qr.; draft as in table; tare 8 .per cent.; duty 15 per cent. ad valorem.

XXV.

RATIO.

RATIO is the mutual relation of two quantities of the same kind to one another.

By finding how many times one number is contained in another, or what part one number is of another, we obtain their ratio. Thus, the ratio of 2 to 4 is 2, because 2 is contained 2 times in 4; and the inverse ratio is $\frac{2}{4}$, because 2 is $\frac{2}{4}$ of 4. Both these expressions of the ratio of 2 to 4 amount to the same thing, which is, that one of the numbers is twice as great as the other.

By the ratio of two quantities is meant only their *relative* magnitude; for, notwithstanding the *absolute* magnitude of 2 pounds and 8 pounds is much greater than that of 2 ounces and 8 ounces, yet the *relative* magnitude or ratio of the two latter is just the same with that of the two former; because, 2 ounces are contained just as many times in 8 ounces, as 2 pounds are in 8 pounds; or, 2 ounces are just as great a part of 8 ounces, as 2 pounds of 8 pounds.

It is evident that only quantities of the same denomination can have a ratio to one another; for it would be absurd to inquire how many times 1 dollar is contained in 4 rods, or what part of 4 rods 1 dollar is.

A ratio is denoted by two dots, similar to a colon: thus, 3 : 9 expresses the ratio of 3 to 9. The former term of a ratio is called the *antecedent*, and the latter the *consequent*. Thus 6 : 12 expresses the ratio of 6 to 12, in which 6 is the antecedent, and 12 the consequent.

Since a ratio indicates how many times one number is contained in another, or what part one number is of another, it is a quotient resulting from the division of one of the terms of the ratio by the other, and may be expressed in the form of a fraction: thus, the ratio 6 : 3 may be expressed by the fraction $\frac{3}{6}$, or conversely $\frac{6}{3}$.

When any two numbers are multiplied, each by the same number, the ratio of the products is the same with the ratio of the multiplicands. Thus, take 3 : 6, and multiply the antecedent and consequent, each by 5, and the products 15 and 30 have the same ratio with 3 and 6; that is, 15 is contained just as many times in 30, as 3 is in 6; or 15 is the same part of 30, that 3 is of 6.

Also, if two numbers be divided, each by the same number, the ratio of the quotients is the same with the ratio of the dividends. Thus, take the ratio of 9 : 18, and divide each term by 3, and the quotients 3 and 6 have the same ratio with 9 and 18; because 3 is contained as many times in 6, as 9 is in 18; or 3 is the same·part of 6, that 9 is of 18.

A ratio resulting from the multiplication of two or more ratios together, that is, the antecedents into the antecedents, and the consequents into the consequents, is called a compound ratio. Thus, 6 : 48 is the compound ratio of 1 : 2, 3 : 4, and 2 : 6; because 6 is the product of all the antecedents, and 48 of all the consequents. This is expressed in fractions with the word "of" between them: thus, making the antecedents the numerators, $\frac{1}{2}$ of $\frac{3}{4}$ of $\frac{2}{6}$; making the consequents the numerators, $\frac{2}{4}$ of $\frac{4}{3}$ of $\frac{6}{2}$.

Two ratios may be equal to one another, as well as two quantities. The equality of two ratios is denoted by ·the sign placed between‚them; thus, 2 : 4 = 3 : 6 signifies that the ratio of 2 to 4 is equal to the ratio of 3 to 6.

PROPORTION.

The equality of 2 ratios is called a **PROPORTION**, and the terms· are called *proportionals;* and in a proportion, the first and fourth terms, that is, the antecedent of the first

ratio and the consequent of the second, are called the
extreme terms; and the second and third terms, that is,
the consequent of the first ratio and the antecedent of
the second, are called the *mean* terms. Thus, in the proportion $3 : 9 = 4 : 12$, 3 and 12 are the extreme terms,
9 and 4 the mean terms.

If the antecedent of the second ratio be the same with
the consequent of the first, the terms are in continued
proportion. Thus, 3, 9, and 27, are in continued proportion, because $3 : 9 = 9 : 27$.

Since the equality of two ratios constitutes a proportion, we can easily decide whether any four numbers be
in proportion, by bringing the fractions expressing the two
ratios to a common denominator; for then, if the numbers
be proportionals, the numerators also will be equal to one
another.

Take the numbers 4, 2, 6, 3; if we make the consequents the numerators, the fraction expressing the ratio
of the two first in the series is $\frac{2}{4}$, and that expressing the
ratio of the two last is $\frac{3}{6}$. These fractions, when reduced
to a common denominator, become $\frac{12}{24}$ and $\frac{12}{24}$; and this
equality of the two fractions expressing the two ratios,
proves that the four numbers are proportionals; for, if
the four numbers were not in proportion, the fraction
expressing the first ratio not being equal to the fraction
expressing the second ratio, the numerator of the one
would not be equal to the numerator of the other, when
reduced to a common denominator.

Again, let us take the same numbers, 4, 2, 6, 3, and
make the antecedents the numerators of the fractions
expressing the ratios: thus, $\frac{4}{2}$ and $\frac{6}{3}$. These fractions when
reduced to a common denominator, are $\frac{12}{6}$ and $\frac{12}{6}$, which,
being equal, prove the four numbers to be proportionals.

We see, therefore, whether we make the antecedents
or consequents the numerators of the fractions expressing
the ratios, that in both cases the equality of the ratios
proves a proportion among the four numbers; and in both
cases the numerators are precisely the same; for in the
first case the fractions are $\frac{12}{24}$ and $\frac{12}{24}$, and in the second,
$\frac{12}{6}$ and $\frac{12}{6}$, and these numerators, in both cases, are

12

products obtained by multiplying together the extreme terms 4 and 3, and the mean terms 2 and 6.

These results prove, that, *if four numbers be in proportion, the product of the two extreme terms is equal to the product of the two mean terms:* a principle of great practical utility, and the foundation of the ancient RULE OF THREE.

It follows from what has been said, that the order of the terms of a proportion may be changed, provided they be so placed, that the product of the extremes shall be equal to the product of the means; because, whenever the product of the extreme terms of four numbers is equal to the product of the mean terms, the numbers are proportionals.

Take, for example, the proportion $\quad$ 3 : 9 = 8 : 24
3 : 9 = 8 : 24, and observe the dif- $\quad$ 3 : 8 = 9 : 24
ferent orders in which its terms may $\quad$ 24 : 8 = 9 : 3
be arranged. $\qquad$ 24 : 8 = 8 : 3

That these changes do not disturb the proportion is evident; for the same numbers, which are the extreme terms in the first proportion, are the extreme terms in all the proportions; and the numbers, which are the mean terms in the first proportion, are the mean terms in all the proportions; therefore the products of the extremes and the products of the means must be the same in all the proportions.

Again, the order of the above pro- $\quad$ 9 : 3 = 24 : 8
portionals may be so changed, that $\quad$ 8 : 3 = 24 : 9
the mean terms shall become the ex- $\quad$ 9 : 24 = 3 : 8
treme terms, and the extreme terms $\quad$ 8 : 24 = 3 : 9
the mean terms.

Since both the terms of a ratio may be multiplied or divided by the same number without altering the ratio, it follows, that all the terms of a proportion may be multiplied or divided by the same number without disturbing the proportion. Let us take, for example, the proportion 2 : 4 = 6 : 12, and multiply each of the terms by 2, and we shall have the proportion 4 : 8 = 12 : 24. If, instead of multiplying, we divide the terms of the same proportion by 2, we shall have the proportion 1 : 2 = 3 : 6.

Either the two antecedents or the two consequents, in two equal ratios, may be multiplied or divided by the same number without destroying the proportion; because the two ratios are increased or diminished alike, and therefore remain equal.   Take the proportion $4 : 16 = 6 : 24$, and multiply each of the antecedents by 2, and it will be $8 : 16 = 12 : 24$; if, instead of the antecedents, we multiply the consequents by 2, we have the proportion $4 : 32 = 6 : 48$; if, instead of multiplying, we divide each of the antecedents by 2, we have the proportion $2 : 16 = 3 : 24$; if, instead of the antecedents, we divide the consequents, we have the proportion $4 : 8 = 6 : 12$.

We may also multiply the antecedents and divide the consequents at the same time, and vice versâ, without destroying the proportion.   If, for example, we take the proportion $3 : 6 = 9 : 18$, and multiply each of the antecedents by 3, and divide each of the consequents by the same number, we have the proportion $9 : 2 = 27 : 6$; if we multiply the antecedents by 3, and divide the consequents by 2, we have the proportion $9 : 3 = 27 : 9$; if we divide the antecedents by 3, and multiply the consequents by 2, we have $1 : 12 = 3 : 36$.

Two or more proportions may be multiplied together, term by term, and the products will be proportionals; for it is the same as multiplying two equal fractions by two other equal fractions, the products of which will again be equal to each other. We give the following as an example.

$$3 : 4 = 6 : 8$$
$$2 : 3 = 8 : 12$$
$$\overline{6 : 12 = 48 : 96}$$

We may also divide one proportion by another, term by term, with equal correctness of conclusion; for this is only dividing two equal fractions by two other equal fractions, the quotients of which will again be equal. Take, for example, the proportion $24 : 32 = 27 : 36$, and divide it by the proportion $6 : 2 = 9 : 3$, term by term, and it gives $4 : 16 = 3 : 12$.

A great variety of other changes may be made by differently multiplying, or dividing, or both; and such changes are frequently convenient in solving questions.

The magnitudes of proportionals are changed without
destroying the proportionality, when either the antecedents, or consequents, or both, are respectively increased or diminished by quantities having the same ratio; or
when the two terms of either or of both ratios are respectively increased or diminished by quantities in the
same ratio with themselves. We will make a few such
changes in the proportionals 8, 6, 20, and 15.

$$8 : 6 \qquad = 20 : 15$$
$$8 + 6 : 6 \qquad = 20 + 15 : 15, \qquad \text{or } 14 : 6 = 35 : 15$$
$$8 - 6 : 6 \qquad = 20 - 15 : 15, \qquad \text{or } 2 : 6 = 5 : 15$$
$$8 : 8 - 6 \qquad = 20 : 20 - 15, \qquad \text{or } 8 : 2 = 20 : 5$$
$$8 : 8 + 6 \qquad = 20 : 20 + 15, \qquad \text{or } 8 : 14 = 20 : 35$$
$$8 + 6 : 8 - 6 = 20 + 15 : 20 - 15, \text{ or } 14 : 2 = 35 : 5$$

Since the product of the extremes in every proportion
is equal to the product of the means, one product may
be taken for the other: now if we divide the product of
the extremes by one extreme, the quotient is the other
extreme; therefore, if we divide the product of the
means by one extreme, the quotient is the other extreme:
for the same reason if we divide the product of the extremes by one of the means, the quotient is the other
mean; consequently, we can find any one term of a proportion, when we know the other three.

To apply these principles to practice, let it be asked—
If 64 yards of cloth cost 304 dollars, what will 36 yards
cost ? In the first place, the ratio of the two pieces of
cloth is 64 : 36; and secondly, the prices are in the same
ratio; that is, 304 dollars must have the same ratio to the
price of 36 yards, that 64 yards have to 36 yards. Now,
if we put A. instead of the answer, we shall have the following proportion, 64 : 36 = 304 : A, in which the product of the means is 10944, which, being divided by 64,
one of the extremes, gives the quotient 171, the other
extreme, which was the term sought; therefore, 171 dollars is the price of 36 yards.

Of the four numbers, which constitute a proportion,
two are of one kind, and two of another. In the preceding example, two of the terms are yards, and two
are dollars.

If there are different denominations in the two first terms, they must both be reduced to the lowest denomination in either of them; and the third term must be reduced to the lowest denomination mentioned in it. Thus, if 4 yards cost 18 shillings and 6 pence, what will 3 yards 1 quarter 2 nails cost? Nails being the lowest denomination in the two first terms, they must both be reduced to nails; pence being the lowest denomination in the third term, this term must be reduced to pence; and when thus reduced, the terms will make the following proportion; $\overset{\text{nai.}}{64} : \overset{\text{nai.}}{54} = \overset{\text{pence}}{222} : \overset{\text{pence}}{A}$. The answer, when obtained, being in pence, must be reduced to shillings and pounds. In this question the answer is 15 s. $7\frac{6}{16}$ d.

From the principles of ratio and proportion, which have been explained, we deduce the following rule for solving questions.

RULE. *Make the number, which is of the same kind with the answer, the third term; and, if from the nature of the question, the third term must be greater than the fourth term or answer, make the greater of the two remaining terms the first term, and the smaller the second; but, if the third term must be less than the fourth, make the less of the two remaining terms the first term, and the greater the second: then multiply the second and third terms together, and divide the product by the first term: the quotient will be the fourth term, or answer.*

1. If I buy 871 yards of cotton cloth for 78 dollars 39 cents, what is the price of 29 yards of the same?

871 : 29 = 78.39 : A

$$29$$

$$\overline{70551}$$
$$15678$$

$$871)2273.31(2.61 \text{ } Ans.$$
$$1742$$
$$\overline{5313}$$
$$5226$$
$$\overline{871}$$
$$871$$

12*

The statements of this question may be read thus —The ratio of 871 to 29 is equal to the ratio of 78.39 to the answer. Or thus—As 871 yd. is to 29 yd., so is $78.39 to the answer. The operation amounts to nothing more than the multiplication of 78.39 by $\frac{39}{871}$

2. If $1\frac{3}{4}$ yard of cotton cloth cost 42 cents, what will $87\frac{1}{2}$ yards cost?

3. If I can buy $1\frac{1}{4}$ yard of cotton cloth for $6\frac{1}{4}$ pence, how many yards can I buy for £10 6 s. 8 d.?

4. If I buy 54 barrels of flour for 297 dollars, what must I give for 73 barrels, at the same rate?

5. If 7 workmen can do a piece of work in 12 days, how many can do the same work in 3 days?

6. If 20 horses eat 70 bushels of oats in 3 weeks, how many bushels will 6 horses eat in the same time?

7. If a piece of cloth containing 76 yards cost 136 dollars 80 cents, what is that per ell English?

8. If a staff 4 feet long cast a shadow 7 feet in length, on level ground, what is the height of a steeple, whose shadow at the same time measures 198 feet?

9. How many yards of paper $2\frac{1}{2}$ feet wide, will hang a room, that is 20 yards in circuit, and 9 feet high?

10. A certain work having been accomplished in 12 days by working 4 hours a day, in what time might it have been done by working 6 hours a day?

11. If 12 gallons of wine are worth 30 dollars, what is the value of a cask of the same kind of wine, containing $31\frac{1}{2}$ gallons?

12. If $8\frac{3}{4}$ yards of cloth cost 4 dollars 20 cents, what will $13\frac{1}{2}$ yards cost, at the same rate?

13. How many yards of cloth $\frac{3}{4}$ yard wide, are equal to 30 yards $1\frac{1}{4}$ yard wide?

14. If 7 pounds of sugar cost 75 cents, how many pounds can I buy for 6 dollars?

15. If 2 pounds of sugar cost 25 cents, and 8 pounds of sugar are worth 5 pounds of coffee, what will 100 pounds of coffee cost?

16. A merchant owning $\frac{4}{5}$ of a vessel, sold $\frac{2}{5}$ of his share for 957 dollars. What was the vessel worth at that rate?

17. A merchant failing in trade, owes 62936 dollars 39 cents; but his property amounts to only 38793 dollars 96 cents, which his creditors agreed to accept, and discharge him. How much does the creditor receive, to whom he owes 2778 dollars 63 cents?

18. Bought 3 tons of oil, for 503 dollars 25 cents; 85 gallons of which having leaked out. I wish to know at what price per gallon I must sell the residue, that I may neither gain nor lose by the bargain.

19. If, when the price of wheat is 6 s. 3 d. a bushel, the penny loaf weighs 9 oz., what ought it to weigh, when wheat is at 8 s. 2½ d. a bushel?

20. If 15 yards of cloth ¾ yard wide cost 6 dollars 25 cents, what will 40 yards being yard wide cost?

21. What quantity of water must I add to a pipe of mountain wine, for which I gave 110 dollars, to reduce the first cost to 75 cents a gallon?

22. Borrowed of a friend 250 dollars for 7 months; and then, to repay him for his kindness, I loaned him 300 dollars. How long must he keep the 300 dollars, to balance the previous favor?

23. If 4½ cwt. can be carried 36 miles for 5¾ dollars, how many pounds can be carried 20 miles for the same money?

24. A person owning ⅔ of a coal mine, sells ¾ of his share for 570 dollars. What is the whole mine worth, at the same rate?

25. If the discount on $106, for a year, be $6, what is the discount on $477, for the same time?

26. If, when the days are 13⅔ hours long, a traveller perform his journey in 35½ days, in how many days will he perform the same journey, when the days shall be 11$\frac{9}{10}$ hours long?

27. A regiment of soldiers consisting of 976 men, is to be new clothed; each coat to contain 2½ yards of cloth 1⅜ yard wide, and to be lined with shalloon ⅞ yard wide. How many yards of shalloon will be required?

28. If 30 men can perform a piece of work in 15 days, how many men would accomplish the same piece of work, in a fifth part of the time?

29. What is the value of 172 pigs of lead, each weighing 3 cwt. 2 qrs. 17½ lb., at 29 dollars 58⅓ cents per fother of 19½ cwt.?

30. A merchant gave his note for 1831 dollars 75 cents, payable in 8 months; but the holder of the note

being pressed for money, the merchant paid it in 3 months. Allowing money to be worth 6 per cent. a year, what sum was requisite to redeem the note?

31. If A can mow an acre of grass in $5\frac{2}{3}$ hours, and B can mow $1\frac{3}{4}$ acre in $9\frac{1}{2}$ hours, in what time can they jointly mow $8\frac{1}{4}$ acres?

32. How much cambrick may be bought for £8 6 s. $3\frac{1}{2}$ d., if 291 yards cost £186 2 s. $4\frac{1}{2}$ d.?

33. If a staff 3 feet 3 inches high casts a shadow $5\frac{1}{2}$ feet long, what is the height of the steeple of Park street church, in Boston, which, at the same time, casts a shadow of 368 feet 6 inches in length?

34. A and B hired a pasture for $49.50, in which, A pastured 13 cows, and B 19. What must each pay?

35. If 220 yards in length and 22 yards in breadth make an acre, what must be the breadth of a lot that is 121 yards in length, to contain an acre?

36. If 365 men consume 75 barrels of provision in 9 months, what number of barrels will 500 men consume during the same time?

37. If 19 yards of linen cost $14.25, what will 435.6 yards come to, at the same rate?

38. The value of 8.25 pounds of pure silver being $128, what is the value of 376.7848 pounds?

39. Suppose sound to move 1106.3 feet in 1 second; how many miles distant is a cloud, in which lightning is observed 47.5 seconds before the thunder is heard?

40. It has been found, that 100 cubic inches of alcohol and 82.5 cubic inches of water, when mixed, measures only 177.41 cubic inches. If, then, 125 gallons of alcohol and 103.125 gallons of water be mixed, how many gallons will the mixture compose?

COMPOUND PROPORTION.

When proportion is applied to questions, in which the relation of the required quantity to the given quantity of the same kind is traced through two or more proportions, it is called COMPOUND PROPORTION.

For example, 16 men dug a trench 54 yards long and 5 yards wide, in 6 days. How many men of equal ability and industry will dig a trench of the same depth, 135 yards long, and 4 yards wide, in 8 days?

In the above question, the number of men required depends upon three circumstances; viz. the length of the trench, its width, and the number of days in which it is to be dug. If we omit the consideration of all the circumstances except the length, the question will be simply this—If 16 men dug a trench 54 yards long, how many men will dig one 135 yards long?—which will make the following proportion; $54 : 135 = 16 : A$, and the fourth term will be found to be 40 men.

Secondly, we will consider the width ; and, since the second trench is to be narrower than the first, the number of men required will be proportionally less, and our second proportion will be the following; $5 : 4 = 40 : A$, and the fourth term will be 32 men.

Lastly, we will notice the number of days in the question ; and, since the longer the time allowed, the less will be the number of men required to do the work, we shall have the following proportion; $8 : 6 = 32 : A$, and this gives 24 men for the fourth term, which is the answer to the question.

We see in this solution, that 16 is multiplied by 135, and the product divided by 54; the quotient, being made the third term in the second proportion, is multiplied by 4, and the product divided by 5; this last quotient, being made the third term in the third proportion, is multiplied by 6, and the product divided by 8. The result, therefore, would be the same, if 135 and 4 and 6 were multiplied together, and their product multiplied by 16, and this last product divided by the product of 54 and 5 and 8. The proportion may be thus arranged.

$$\left.\begin{array}{r} 54 : 135 \\ 5 : \quad 4 \\ 8 : \quad 6 \end{array}\right\} = 16 : A$$

$$\overline{2160 : 3240} = 16 : A$$

If, instead of calculating the fourth term in each proportion, we only indicate the operation by a fraction, we shall have, in the first of the foregoing proportions, $\frac{16 \times 135}{54}$ for the fourth term: taking this for the third term of the second proportion, we shall have the following, $5 : 4 = \frac{16 \times 135}{54} : A$, and the fourth term will be $\frac{16 \times 135 \times 4}{54 \times 5}$; taking this for the third term in the third proportion, we shall have the following, $8 : 6 = \frac{16 \times 135 \times 4}{54 \times 5} : A$, and the fourth term will be $\frac{16 \times 135 \times 4 \times 6}{54 \times 5 \times 8}$, which is equal to 24, the number of men required. In this fractional expression, we see at once, that the product of all the second terms is multiplied by the third term, and that this product is divided by the product of all the first terms, and the quotient is the fourth term, or answer to the question.

Hence we see, that questions in compound proportion will be accurately solved by the following rule.

RULE. *Make the number, which is of the same kind with the answer, the third term; of the remaining numbers, take any two of a kind, and write one for a first term and the other for a second term, as directed in simple proportion, then any other two of a kind, and so on, till all are written. Lastly, multiply all the second terms together, and their product by the third term, and divide the result by the product of the first terms; the quotient will be the fourth term, or answer.*

41. A wall to be built to the height of 27 feet, was raised to the height of 9 feet by 12 men in 6 days; how many men must be employed to finish it in 4 days?

$$\begin{array}{r} 27 \\ 9 \\ \hline \end{array}$$

$\left. \begin{array}{r} 9 : 18 \\ 4 : 6 \end{array} \right\} = 12 : A$

$36 : 108 \quad = 12 : A$

$\begin{array}{r} 12 \\ \hline \end{array}$

$36)1296(36 \; Ans.$
$\underline{108}$
$\;\;216$
$\;\;\underline{216}$

12 men is the third term, because the answer is to be in men.

In stating the different lengths of the wall, the shorter is made the first term, because, the longer the wall, the greater will be the number of men required to build it.

In stating the days, the less number is made the first term, because, the less the time, the greater must be the number of men.

42. If 120 bushels of corn will serve 14 horses 56 days, how many days will 94 bushels serve 6 horses?

43. If a footman travel 130 miles in 3 days, when the days are 12 hours long, in how many days of ten hours in length, can he travel 360 miles?

44. If 6 laborers dig a ditch 34 yards long in 10 days, how many yards will 20 laborers dig in 15 days?

45. If a garrison of 600 men have provisions for 5 weeks, allowing each man 12 ounces a day, how many men may be maintained 10 weeks with the same provisions, if each man is limited to 8 ounces a day?

46. If 3 bushels and 3 pecks of wheat will last a family of 9 persons 22 days, in how many days will 6 persons consume 5 bushels?

47. If 450 tiles, each 12 inches square, will pave my cellar, how many tiles must I have, if they are only 9 inches long and 8 inches broad?

48. If 12 ounces of wool make $2\frac{1}{2}$ yards of cloth 6 quarters wide, how much wool is required for 150 yards 4 quarters wide?

49. If a bar of iron 4 feet long, 3 inches broad, and $1\frac{1}{2}$ inch thick, weighs 36 pounds, what will a bar weigh, that is 6 ft. long, 4 in. broad, and 2 in. thick?

50. If 6 men built a wall 20 feet long, 6 feet high, and 4 feet thick, in 16 days, in how many days will 24 men build a wall 200 feet long, 8 feet high, and 6 feet thick?

51. If 14 men can reap 84 acres in 6 days, how many men must be employed to reap 44 acres in 4 days?

52. A ship's crew of 300 men were so supplied with provisions for 12 months, that each man was allowed 30 ounces a day; but after having been 6 months on their voyage, they find it will take 9 months more to finish it, and 50 of their number have been lost. It is required, to find the daily allowance of each man during the last 9 months.

53. A wall was to be built 700 yards long in 29 days; after 12 men had been employed on it for 11 days, it was found they had built only 220 yards. How many more men must be put on, to finish it in the given time?

54. If the transportation of 12 cwt. 2 qr. 6 lb., 275 miles cost $27.78, how far, at that rate, may 3 tons 0 cwt. 3 qr. be carried, for $234.78?

55. A cistern 17½ feet in length, 10½ feet in breadth, and 13 feet deep, holds 546 barrels of water. Then how many barrels will fill a cistern, that is 16 feet long, 7 feet broad, and 15 feet deep?

56. If 25 pears can be bought for 10 lemons, and 28 lemons for 18 pomegranates, and 1 pomegranate for 48 almonds, and 50 almonds for 70 chestnuts, and 108 chestnuts for 2½ cents, how many pears can I buy for $1.35?

57. In how many days, working 9 hours a day, will 24 men dig a trench 420 yards long, 5 yards wide, and 3 yards deep; if 248 men, working 11 hours a day, in 5 days dug a trench 230 yards long, 3 yards wide, and 2 yards deep?

58. If the interest on 347 dollars for 3½ years be 72 dollars 87 cents, what will be the interest, at the same rate, on 537 dollars for 2½ years?

59. What must be paid for the carriage of 4 cwt., 32 miles, if the carriage of 8 cwt., 128 miles, cost 12 dollars 80 cents?

60. By working 9 hours a day, 5 men hoed 18 acres of corn in 4 days. How many acres will 9 men hoe, at that rate, in 3 days, working 10 hours a day?

61. One pound of thread makes 2 yards of linen cloth, 5 quarters wide. Then how many pounds of thread will be required to make 50 yards of linen ¾ yd. wide?

62. If 6 men, working 7 hours a day, mowed 28 acres of grass in 4 days, how many men, at that rate, will mow 16 acres in 8 days, working 6 hours a day?

63. If 5 men can make 300 pair of boots in 40 days, how many men must be employed to make 900 pair in 60 days?

64. If 3 compositors set 15½ pages in 2⅞ days, how many will be required to set 69¾ pages in 6¼ days?

65. If 36 yards of cloth, 7 quarters wide be worth $98, what is the value of 120 yd. of cloth of equal texture, but only 5 quarters wide?

XXVI.

CONJOINED PROPORTION.

CONJOINED PROPORTION—called by merchants, *The Chain Rule*—consists of a series of terms bearing a certain proportion to each other, and so connected, that a comparison is instituted between two of the terms, through the medium of all the others.

The principles of this rule are included in proportion. The rule is chiefly employed in the higher operations of exchange, arbitrations of bullion, specie and merchandise. For the purpose of elucidation, however, we propose the following familiar example.

If 3 lb. of tea be worth 4 lb. of coffee, and 6 lb. of coffee worth 20 lb. of sugar, how many pounds of sugar may be had for 9 lb. of tea?

This question, we know, may be solved by a statement in compound proportion; but the following is the solution by conjoined proportion.

Distinguish the several terms into *antecedents* and *consequents*, and connect them by the sign of equality in the way of equations, as follows.

First, enter on the right the given sum or term on which the operation is to be performed, (which in the foregoing question is 9 lb. of tea) and call this the *term of demand*.

Secondly, on the left of this term, and a line lower, enter the first antecedent, which must be of the same kind or name with the term of demand, and equal in value to the annexed consequent.

Thirdly, in the same manner, let the second antecedent be of the same name with the second consequent, and equal in value to the third consequent: and so on, for any number of terms.

Fourthly, the terms being thus arranged, divide the product of the consequents by the product of the antecedents, and the quotient will be the answer in the denomination of the last consequent, or *odd term*.

13

$$9\,\text{lb. of tea, term of demand.}$$
$$3\,\text{lb. tea} = 4\,\text{lb. of coffee.}$$
$$6\,\text{lb. coffee} = 20\,\text{lb. sugar, the odd term.}$$

Hence $\dfrac{20 \times 4 \times 9}{6 \times 3} = \dfrac{720}{18} = 40$ lb. sugar, the answer.

By the above example it will be seen, that in the arrangement of antecedents and consequents, each sort is entered twice, except that in which the answer is required, and which is called the odd term.

It should also be observed, that no two entries of the same denomination are in the same column; and, as they are placed in the way of equations, it is evident that the quantities on each side, which are equal in value to one another, are cancelled in the operation; and, therefore, the quotient or answer will obviously be in the denomination of the last consequent, which is the odd term.

This rule may be proved by reversing the operation; taking the answer as the term of demand, and making the first antecedent the last consequent or odd term, as follows.

$$40\,\text{lb. of sugar.}$$
$$20\,\text{lb. sugar} = 6\,\text{lb. coffee.}$$
$$4\,\text{lb. coffee} = 3\,\text{lb. tea.}$$

Then $\dfrac{3 \times 6 \times 40}{4 \times 20} = \dfrac{720}{80} = 9$ lb. of tea, the proof.

The operation may be abridged by omitting such numbers as are the same in both columns, whenever such instances exist.

When fractions occur, the most convenient method is to convert them into whole numbers. Thus, an antecedent of $\frac{7}{12}$ and a consequent of 9 may be changed (by multiplying both by 12) into 7 and 108, and the ratio will not be altered. So 5 and $11\frac{3}{4}$ have the same ratio with 20 and 47.

The rule may be exemplified by a question in reduction; thus,— It is required to reduce 2 tons to ounces.

$$2\,\text{tons, term of demand.}$$
$$1\,\text{ton} = 20\,\text{cwt.}$$
$$1\,\text{cwt.} = 4\,\text{qr.}$$
$$1\,\text{qr.} = 28\,\text{lb.}$$
$$1\,\text{lb.} = 16\,\text{oz. the odd term.}$$

Then $\dfrac{16 \times 28 \times 4 \times 20 \times 2}{1 \times 1 \times 1 \times 1} = 71680$ ounces, the answer.

1. If 17 lb. of raisins are worth 20 lb. of almonds, and 5 lb. of almonds worth 8½ lb. of figs, and 37½ lb. of figs worth 30 lb. of tamarinds, how many pounds of tamarinds are equal in value to 42½ lb. of raisins?

2. Suppose 100 lb. of Venice weight is equal to 70 lb. of Lyons, and 60 lb. of Lyons to 50 lb. of Rouen, and 20 lb. of Rouen to 25 lb. of Toulouse, and 50 lb. of Toulouse to 37 lb. of Geneva; then how many pounds of Geneva are equal to 25 lb. of Venice?

3. If 1 French crown is equal in value to 80 pence of Holland, and 83 pence of Holland to 48 pence English, and 40 pence English to 70 pence of Hamburgh, and 64 pence of Hamburgh to 1 florin of Frankfort, how many florins of Frankfort are equal to 166 French crowns?

4. If A can do as much work in 3 days as B can do in 4½ days, and B as much in 9 days as C in 12 days, and C as much in 10 days as D in 8 days, how many days' work of D are equal to 5 days' work of A?

5. If 70 braces at Venice are equal to 75 braces at Leghorn, and 7 braces at Leghorn are equal to 4 yards in the United States, how many braces at Venice are equal to 64 yards in the United States?

6. A merchant in St. Petersburg owes 1000 ducats in Berlin, which he wishes to pay in rubles by the way of Holland; and he has for the data of his operation, the following information, viz. That 1 ruble gives 47½ stivers; that 20 stivers make 1 florin; 2½ florins 1 rix dollar of Holland; that 100 rix dollars of Holland fetch 142 rix dollars of Prussia; and that 1 ducat in Berlin is worth 3 rix dollars Prussian. How many rubles will pay the debt?

7. If 94 piasters at Leghorn are equal to 100 ducats at Venice, and 1 ducat is equal to 320 maravedis at Cadis, and 272 maravedis are equal to 630 reas at Lisbon, and 400 reas are equal to 50 d. at Amsterdam, and 56 d. are equal to 3 francs at Paris; and 9 francs are equal to 94 d. at London, and 54 d. are equal to 1 dollar n the United States, how many dollars are equal to 800 piasters?

XXVII.

DUODECIMALS.

DUODECIMALS are compound numbers, the value of whose denominations diminish in a uniform ratio of 12. They are applied to square and cubic measure.

The denominations of duodecimals are the foot, $(f.)$, the prime or inch, $(')$, the second, $('')$, the third, $(''')$, the fourth, $('''')$, the fifth, $(''''')$, and so on. Accordingly, the expression 3 1' 7'' 9''' 6'''' denotes 3 feet 1 prime 7 seconds 9 thirds 6 fourths.

The accents, used to distinguish the denominations below feet, are called *indices*.

The foot being viewed as the unit, duodecimals present the following relations.

$1' = \frac{1}{12}$ of 1 foot.

$1'' = \frac{1}{12}$ of $\frac{1}{12}$ of 1 foot.   . . . $= \frac{1}{144}$ of 1 foot.

$1''' = \frac{1}{12}$ of $\frac{1}{12}$ of $\frac{1}{12}$ of 1 foot. . . $= \frac{1}{1728}$ of 1 foot.

$1'''' = \frac{1}{12}$ of $\frac{1}{12}$ of $\frac{1}{12}$ of $\frac{1}{12}$ of 1 foot. $= \frac{1}{20736}$ of 1 foot.
             &c.

Addition and subtraction of duodecimals are performed as addition and subtraction of other compound numbers; 12 of a lower denomination making 1 of a higher. Multiplication, however, when both the factors are duodecimals, is peculiar, and will now be considered.

When feet are multiplied by feet, the product is in feet. For instance, if required to ascertain the superficial feet in a board 6 feet long and 2 feet wide, we multiply the length by the breadth, and thus find its superficial, or square feet to be 12. But when feet are multiplied by any number of inches [primes], the effect is the same as that of multiplying by so many twelfths of a foot, and therefore the product is in twelfths of a foot, or inches: thus a board 6 feet long and 6 inches wide contains 36 inches, because the length being multiplied by the breadth, that is, 6 feet by $\frac{6}{12}$ of a foot, the product is $\frac{36}{12}$ of a foot, or $36' = 3$ feet. When feet are multiplied by seconds, the product is in seconds: thus 6 feet multiplied by 6 seconds, that is, $\frac{6}{1}$ of a foot by, $\frac{6}{12}$ of $\frac{1}{12}$ of a foot, the product is $\frac{36}{144}$ of a foot, or $36'' = 3$ inches.

When inches are multiplied by inches, the product is in seconds. Thus, 6 inches multiplied by 8 inches, that is, $\frac{6}{12}$ of a foot by $\frac{8}{12}$ of a foot, the product is $\frac{48}{144}$ of a foot, or $48'' = 4$ inches. When inches are multiplied by seconds, the product is in thirds. Thus, 6 inches multiplied by 8 seconds, that is, $\frac{6}{12}$ of a foot by $\frac{8}{12}$ of $\frac{1}{12}$ of a foot, the product is $\frac{48}{1728}$ of a foot, or $48''' = 4$ seconds. When seconds are multiplied by seconds, the product is in fourths. Thus, $6''$ multiplied by $8''$, that is, $\frac{6}{12}$ of $\frac{1}{12}$ of a foot, by $\frac{8}{12}$ of $\frac{1}{12}$ of a foot, the product is $\frac{48}{20736}$ of a foot, or $48'''' = 4$ thirds.

This method of showing the denomination of the product resulting from the multiplication of duodecimals by duodecimals may be extended to any number of places whatever; but sufficient has been said, to show that the product is always of that denomination denoted by the sum of the indices of the two factors.

Feet multiplied by feet, produce feet.
Feet multiplied by primes, produce primes.
Feet multiplied by seconds, produce seconds.
Feet multiplied by thirds, produce thirds.
&c.
Primes multiplied by primes, produce seconds.
Primes multiplied by seconds, produce thirds.
Primes multiplied by thirds, produce fourths.
&c.
Seconds multiplied by seconds, produce fourths.
Seconds multiplied by thirds, poduce fifths.
Seconds multiplied by fourths, produce sixths.
&c.
Thirds multiplied by thirds, produce sixths.
Thirds multiplied by fourths, produce sevenths.
&c.

If we would find the square feet in a floor 6 f. 4′ 8″ in length, and 4 f. 6′ 5″ in breadth, we should proceed as follows.

13*

$$6\text{f.}\quad 4'\quad 8''$$
$$4\quad 6'\quad 5''$$

$$\overline{2'\ \ 7''\ 11'''\ 4''''}$$
$$3\quad 2'\quad 4''\quad 0'''$$
$$25\quad 6'\quad 8''$$

$$\overline{28\text{ f.}\ 11'\ 7''\ 11'''\ 4''''}$$

We began on the right hand, and multiplied the whole multiplicand, first by the seconds in the multiplier, then by the inches, and lastly by the feet, and added the results together, and thus obtained the answer.

That the above answer is the true one, will appear very clearly from the following considerations. The 8 seconds, as we have already shown, may be considered in relation to feet as $\frac{8}{144}$, and the 5 seconds as $\frac{5}{144}$, the product of which is $\frac{40}{20736}$ of a foot, or 40'''', which is equal to 3''' and 4''''; writing down the 4'''', we reserve the 3''' to be added to the product of 4' by 5". 4' being $\frac{4}{12}$ of a foot, and 5" being $\frac{5}{144}$ of a foot, their product is $\frac{20}{1728}$ of a foot, or 20''', to which adding the 3''', that were reserved, we had 23''', equal to 1" and 11'''; we wrote down the 11''', and reserved the 1" to be added to the product of 6 feet by 5". 6 feet being $\frac{6}{1}$ of a foot, and 5" being $\frac{5}{144}$ of a foot, their product is $\frac{30}{144}$ of a foot, or 30", to which we added the 1" reserved, and thus had 31", equal to 2' and 7", both of which we wrote down.

Having completed the multiplication by the seconds, we next multiplied by the inches: 8" being $\frac{8}{144}$ of a foot, and 6' being $\frac{6}{12}$ of a foot, their product is $\frac{48}{1728}$ of a foot, or 48''' = 4"; we therefore put down a cipher in the place of thirds, and reserved the 4" to be added to the product of inches by inches. 4 inches being $\frac{4}{12}$ of a foot, and 6 inches $\frac{6}{12}$ of a foot, their product is $\frac{24}{144}$ of a foot, or 24", to which we added the 4" reserved, making 28" = 2' and 4"; writing down the 4", we reserved the 2' to be added to the product of feet by inches. 6 feet being $\frac{6}{1}$ of a foot, and 6 inches $\frac{6}{12}$ of a foot, their product is $\frac{36}{12}$ of a foot, or 36', to which we added the 2' reserved, making 38' = 3 feet and 2 inches, both of which we wrote down.

Lastly, we multiplied by the feet in the multiplier. 8", or $\frac{8}{144}$ of a foot being multiplied by 4 feet, or $\frac{4}{1}$ of a foot, their product is $\frac{32}{144}$ of a foot, or 32" = 2' and 8"; setting down the 8", we reserved the 2' to be added to the product of inches by feet. 4', or $\frac{4}{12}$ of a foot being multiplied by 4 feet, or $\frac{1}{4}$ of a foot, their product is $\frac{16}{12}$ of a foot, or 16', to which we added the 2' reserved, making 18'=1 foot and 6 inches; writing down the 6', we reserved the 1 foot to be added to the product of feet by feet. 6 feet being multiplied by 4 feet, their product is 24 feet, to which we added the 1 foot reserved, making 25 feet. By adding these three partial products together, we obtained the answer to the question.

Therefore, to multiply one number consisting of feet, inches, seconds, &c. by another of the same kind, we give the following rule.

RULE. *Place the several terms of the multiplier under the corresponding ones of the multiplicand. Beginning on the right hand, multiply the several terms of the multiplicand by the several terms of the multiplier successively, placing the right hand term of each of the partial products under its multiplier; then add the partial products together, observing to carry one for every twelve, both in multiplying and adding. The sum of the partial products will be the answer.*

Questions in duodecimals are very commonly performed by commencing the multiplication with the highest denomination of the multiplier, and placing the partial products as in the first of the two following operations. The result is the same, whichever method is adopted. The second operation, however, is according to the rule we have given, and is more conformable to the multiplication of numbers accompanied by decimals.

3 f.	2'	7″			3 f.	2'	7″	
2 f.	6'	4″			2 f.	6'	4″	
6	5'	2″			1'	0″	10‴	4⁗
1	7'	3″	6‴		1	7'	3″	6‴
	1'	0″	10‴	4⁗	6	5'	2″	
8 f.	1'	6″	4‴	4⁗	8 f.	1'	6″	4‴ 4⁗

When there are not feet in both the factors, there may not be any feet in the product; but, after what has been said, there will be no difficulty in determining the places of the product.

1. Multiply 14 f. 9′ by 4 f. 6′.

2. What are the contents of a marble slab, whose length is 5 f. 7′, and breadth 1 f. 10′?

3. How many square feet are there in the floor of a hall 48 f. 6′ long, and 24 f. 3′ wide?

4. Multiply 4 f. 7′ 8″ by 9 f. 6′.

5. How many square feet are there in a house lot 43 f. 3′ in length, and 25 f. 6′ in breadth?

6. What is the product of 10 f. 4′ 5″ by 7 f. 8′ 6″?

7. Calculate the square feet in an alley 44 f. 2′ 9″ long, and 2 f. 10′ 3″ 2‴ 4⁗ wide.

8. How many square feet are there in a garden 39 f. 10′ 7″ long, and 18 f. 8′ 4″ wide?

9. What is the product of 24 f. 10′ 8″ 7‴ 5⁗ by 9 f. 4′ 6″?

10. Compute the solid feet in a wall 53 f. 6′ long, 12 f. 3′ high, and 2 f. thick.

11. The length of a room is 20 feet, its breadth 14 feet 6′, and its height 10 f. 4′. How many yards of painting are there in its walls, deducting a fire place of 4 f. by 4 f. 4′; and two windows, each 6 f. by 3 f. 2′?

12. How many solid feet in a pile of wood 22 f. 6′ long, 12 f. 8′ wide, and 5 f. 8′ high?

13. How many yards of plastering in the top and four walls of a hall 58 f. 8′ long, 21 f. 4′ wide, and 13 f. 9′ high; deducting for 2 doors each 7 f. 6′ high and 4 f. wide; for 7 windows each 6 f. 2′ high, and 3 f. 10′ wide; for 2 fire places, each 3 f. 6′ high, and 4 f. wide, and for a mop board 9 inches wide around the hall?

14. How many yards of papering in a room 17 f. 8′ long, 16 f. 9′ wide, and 12 f. 6′ high; deducting for 2 doors each 6 f. 6′ high, and 4 f. wide; for a fire place 4 f. 6′ high and 3 f. 10′ wide; for 3 windows each 5 f. 6′ high and 3 f. 8′ wide, and for a mop board 8 inches wide around the room?

15. How many yards of carpeting, yard wide, will be required for a room 21 f. 6′ long, and 18 ft. wide?

16. What will the plastering of a ceiling come to, at 10 cents a square yard, supposing the length 21 feet 8 inches, and the breadth 14 feet 10 inches?

XXVIII.

INVOLUTION.

INVOLUTION is the multiplication of a number by itself. The number, which is thus multiplied by itself is·called the root. The product, which we obtain by multiplying a number by itself, is called a power of that number. The number involved is itself the first power, and is the root of all the other powers.

A number, multiplied once by itself, is said to be involved or raised to the second degree, or second power; multiplied again, to the third degree, or third power; and so on. For example, 3×3 is raised to the second power of ·3, which is 9; $3 \times 3 \times 3$ is raised to the third power of 3, which is 27; &c.

We distinguish the powers from one another by the number of times, that the root is used as factor in the multiplication of itself. Thus, 3×3 produces the second power of 3, because 3 is used twice as factor; $3 \times 3 \times 3$ produces the third power of 3, because 3 is used three times as factor; $3 \times 3 \times 3 \times 3$ produces the fourth power of 3, because 3 is used four times as factor; and so on.

A fraction is involved in the same manner by multiplying it continually into itself; thus, the second power of $\frac{3}{4}$, is $\frac{3}{4} \times \frac{3}{4} = \frac{9}{16}$; the third power is $\frac{9}{16} \times \frac{3}{4} = \frac{27}{64}$; the fourth power is $\frac{27}{64} \times \frac{3}{4} = \frac{81}{256}$; and so on. So also in decimals the second power of .2, is $.2 \times .2 = .04$; the third power is $.04 \times .2 = .008$; the fourth power is $.008 \times .2 = .0016$; and so on.

To involve a mixed number, reduce it first to an improper fraction, or the vulgar fraction to a decimal, and then involve it. Thus, $1\frac{1}{2}$ when reduced to an improper fraction, is $\frac{3}{2}$, the second power of which $\frac{9}{4} = 2\frac{1}{4}$; the third power is $\frac{27}{8} = 3\frac{3}{8}$; &c. If, instead of reducing $1\frac{1}{2}$ to an improper fraction, we reduce the vulgar fraction to a decimal, we have 1.5, the second power of which is 2.25; the third power, 3.375; &c.

The second power is commonly called the square; the third power, the cube; the fourth power, the biquadrate. The other powers now generally receive no other than numeral distinctions; as the fifth power, the sixth power, the seventh power, &c. In some books, however, the fifth power is called the first sursolid; the sixth power, the square cubed, or the cube squared; the seventh power, the second sursolid; the eighth power, the biquadrate squared; the ninth power, the cube cubed.

The powers of 1 remain always the same; because, whatever number of times we multiply 1 by itself, the product is always 1.

A power is sometimes denoted by a number, placed at the right hand of the upper part of the root; thus, 5^2 denotes the second power of 5, which is 25; 4^3 denotes the third power of 4, which is 64; 9^4 denotes the fourth power of 9, which 6561; &c. The number, thus used to denote the power, is sometimes called the *exponent* and sometimes the *index*. But the use of these exponents or indices in arithmetic is very limited; they belong chiefly to algebra.

We will now make a few observations on the result arising from the multiplication or division of one power by another. To illustrate this subject, we will take the number 3; we must here observe, however, that since every number is the first power of itself, the exponent 1 is never expressed; so that 3 and 3^1 mean the same thing; the exponent 1 being always understood, when no exponent is expressed. Now 3 multiplied by 3 produces the second power of 3, which may be thus expressed, $3^1 \times 3^1 = 3^2$; so also $3^2 \times 3^1 = 3^3$; and $3^3 \times 3^1 = 3^4$, &c. We have here expressed the exponent 1 for the purpose of showing that we obtain the exponent of the product or power produced, by adding together the exponents of the factors or powers used in producing it. Hence the second power of any number multiplied by the second power of the same number produces the fourth power of that number; thus, $3^2 \times 3^2 = 3^4$: the third power multiplied by the third power gives the sixth power; as $2^3 \times 2^3 = 2^6$: the fourth power multiplied by the second power gives the sixth power; as $2^4 \times 2^2 = 2^6$: the fourth power multiplied by the fourth power produces the eighth power; as $3^4 \times 3^4 = 3^8$: the third power multiplied by the third power, and the product again by the third power gives the ninth power; as $2^3 \times 2^3 \times 2^3 = 2^9$.

Division being the reverse of multiplication, it is evident, that if we subtract the exponent of the divisor from the exponent of the dividend, the remainder is the exponent of the quotient. For example, if we divide the fifth power by the third power, the quotient is the second power; as $3^5 \div 3^3 = 3^2$: if we divide the ninth power by the sixth power, the quotient is the third power; as $6^9 \div 6^6 = 6^3$: if we divide the ninth power by the eighth power, the quotient is the first power; as $6^9 \div 6^8 = 6$.

1. What is the third power of 12?
2. Find the fourth power of 11.
3. Raise 13 to the fifth power.
4. What is the square of 27?
5. What is the square of .27?
6. Raise .7 to the fourth power.
7. What is the eighth power of .2 ?
8. What is the third power of .1 ?
9. What is the square of $\frac{3}{7}$?
10. What is the cube of $\frac{2}{3}$?
11. Involve $\frac{4}{37}$ to the third power.
12. Raise $\frac{3}{7}$ to the fourth power.
13. What is the square of $30\frac{1}{4}$?
14. What is the biquadrate of $3\frac{1}{4}$?
15. Involve 1.1 to the fifth power.
16. Raise $20\frac{1}{2}$ to the fourth power.
17. Raise 8.2 to the third power.
18. What is the fourth power of 17?
19. Divide 7^5 by 7^3, and write the quotient.
20. Multiply 8^2 by 8, and write the product.
21. What is the quotient resulting from $5^7 \div 5^4$?
22. What is the product resulting from $6^3 \times 6$?
23. Multiply 9^2 by 9, and write the product.
24. Divide 4^{10} by 4^6, and write the quotient.
25. What quotient results from $19^9 \div 19^7$?

TABLE OF ROOTS AND POWERS

Roots.	1	2	3	4	5	6	7	8	9
2d. Pow.	1	4	9	16	25	36	49	64	81
3d. Pow.	1	8	27	64	125	216	343	512	729
4th. Pow.	1	16	81	256	625	1296	2401	4096	6561
5th. Pow.	1	32	243	1024	3125	7776	16807	32768	59049
6th. Pow.	1	64	729	4096	15625	46656	117649	262144	531441
7th. Pow.	1	128	2187	16384	78125	279936	823543	2097152	4782969
8th. Pow.	1	256	6561	65536	390625	1679616	5764801	16777216	43046721
9th. Pow.	1	512	19683	262144	1953125	10077696	40353607	134217728	387420489
10th. Pow.	1	1024	59049	1048576	9765625	60466176	282475249	1073741824	3486784401

EVOLUTION.

EVOLUTION is the reverse of involution; for in involution we have the root given, to find the power; but in evolution we have the power given, to find the root.

Power and root are correlative terms; for, as 4 is the square of 2, 2 is the square root of 4; as 8 is the cube of 2, 2 is the cube root of 8; as 16 is the biquadrate of 2, 2 is the biquadrate root of 16; as 32 is the fifth power of 2, 2 is the fifth root of 32: &c.

The extraction of the root is finding a number, which being multiplied into itself the requisite number of times, will reproduce the given number: for example, if we extract the square root of 81, we find it to be 9, because $9 \times 9 = 81$; but if we extract the biquadrate root of 81, we find it to be 3, because $3 \times 3 \times 3 \times 3 = 81$.

Hence the root is designated by the number of times it is used as factor in producing the corresponding power. It is used twice in producing the second power, and is called the second root, or square root: it is used three times in producing the third power, and is called the third root, or cube root: it is used four times in producing the fourth power, and is called the fourth root, or biquadrate root: it is used five times in producing the fifth power, and is called the fifth root: &c.

A number, whose root can be exactly extracted, is called a perfect power, and its root is called a rational number. For example, 4 is a perfect power of the second degree, and 2, its square root, is a rational number; 27 is a perfect power of the third degree, and 3, its cube root, is a rational number; 64 is a perfect power of the second, third, and sixth degrees, and 8 its square root, 4 its cube root, and 2 its sixth root, are rational numbers; $\frac{8}{27}$ is a perfect power of the third degree, and $\frac{2}{3}$, its cube root, is a rational number; .25 is a perfect power of the second degree, and .5, its square root, is a rational number.

In short, any number, which is the exact root of any power, is a rational number, and its power a perfect

14

power: and since any number may be the root of its corresponding power, it follows that any root, which can be exactly expressed by figures, is a rational number.

But there are numbers, whose roots can never be exactly extracted, and these numbers are called imperfect powers, and their roots are called irrational numbers, or *surds*. For example, 2 is not only an imperfect power of the second degree, but an imperfect power of any degree, and not only its square root, but the root in every degree is irrational, or a surd; because no number, either whole or fractional, can be found, which, being involved to any degree, will produce 2. The same is true of many other numbers. In these cases, by using decimals, we can approximate, or come very near to the root, which is sufficient for most purposes. Thus, we find the square root of 2 to be $1.414 +$. The decimal may be carried to any number of places.

Some numbers are perfect powers of one degree, and imperfect powers of another degree. For example, 4 is a perfect power of the second degree, and its square root, which is 2, is rational; but an imperfect power of the third degree, and its cube root, which is $1.587 +$, is a surd: 8 is an imperfect power of the second degree, and its square root, which is $2.828 +$, is a surd; but a perfect power of the third degree, and its cube root, which is 2, is rational: 16 is a perfect power of the second and fourth degrees, and its square root, which is 4, and its biquadrate root, which is 2, are both rational; but an imperfect power of the third degree, and its cube root, which is $2.519 +$, is a surd.

These irrational numbers or surds occur, whenever we endeavor to find a root of any number, which is not a perfect corresponding power; and, although they cannot be expressed by numbers either whole or fractional, they are nevertheless magnitudes, of which we may form an accurate idea. For however concealed the square root of 2, for example, may appear, we know, that it must be a number, which, when multiplied by itself, will exactly produce 2. This property is sufficient to give us an idea of the number, and we can approximate it continually by the aid of decimals.

A radical sign, written thus $\sqrt{}$, and read square root, is used to express the square root of any number, before which it is placed. The same sign with the index of the root written over it, is used to express the other roots· thus $\sqrt[3]{}$ cube root: $\sqrt[4]{}$ biquadrate root: $\sqrt[5]{}$ fifth root: &c. We will give the following radical expressions; $\sqrt{9}=3$ $\sqrt[3]{8}=2$; $\sqrt[4]{81}=3$; $\sqrt[5]{32}=2$; these expressions are read thus; the square root of 9 is equal to 3; the cube root of 8 is equal to 2; the fourth root of 81 is equal to 3; the fifth root of 32 is equal to 2. Hence it is evident that $\sqrt{9}\times\sqrt{9}=9$; $\sqrt[3]{8}\times\sqrt[3]{8}\times\sqrt[3]{8}=8$; &c.

The explanation, which we have given of irrational numbers or surds, will readily enable us to apply to them the known methods of calculation. We know that the square root of 3 multiplied by itself must produce 3, which may be thus expressed, $\sqrt{3}\times\sqrt{3}=3$; also $\sqrt[3]{3}\times\sqrt[3]{3}\times\sqrt[3]{3}=3$; $\sqrt{\frac{3}{7}}\times\sqrt{\frac{3}{7}}=\frac{3}{7}$; $\sqrt{.5}\times\sqrt{.5}=.5$; $\sqrt[3]{\frac{2}{3}}\times\sqrt[3]{\frac{2}{3}}\times\sqrt[3]{\frac{2}{3}}=\frac{2}{3}$; $\sqrt[4]{5}\times\sqrt[4]{5}\times\sqrt[4]{5}\times\sqrt[4]{5}=5$; $\sqrt{2}\times\sqrt{2}=2$.

Instead of the radical sign, a fractional exponent is also used to express the roots of numbers. The numerator indicates the power of the number, and the denominator the root. Thus, $4^{\frac{1}{2}}$ expresses the square root of 4^1, or 4; $4^{\frac{1}{3}}$, the cube root of 4; $4^{\frac{1}{4}}$, the biquadrate root of 4; $4^{\frac{1}{5}}$, the fifth root of 4; $8^{\frac{2}{3}}$, the cube root of the second power of 8; and since the second power of 8 is 64, and the cube root of 64 is 4, the expression $8^{\frac{2}{3}}$ is equal to 4. The expression $4^{\frac{3}{2}}=8$, is read thus, the square root of the third power of 4 is equal to 8. The expression $9^{\frac{1}{2}}$ is equivalent to $\sqrt{9}$: and $8^{\frac{3}{3}}$ is equivalent to $\sqrt{8^3}$: also $4^{\frac{2}{4}}$ is equivalent to $\sqrt[4]{4^2}$: the expression $4^{\frac{2}{4}}$ is also equal to $4^{\frac{1}{2}}$, because $\frac{2}{4}$ is equal to $\frac{1}{2}$.

A line, or vinculum, drawn over several numbers, signifies that the numbers under it are to be considered jointly: thus, $\sqrt{25+11}$ is equal to 6, because $25+11$ is

36, and the square root of 36 is 6; but $\sqrt{25+11}$ is equal to 16, because the square root of 25 is 5, and $5+11$ is 16. The expression $\sqrt{27-6+43}$ is equivalent to $\sqrt{64}$. And $\sqrt[3]{100-73}=3$. Also $20-\sqrt{9+7+1}=17$. Likewise $\sqrt{90-9-4}+\sqrt[3]{53-45}+6=13$.

XXIX.

EXTRACTION OF THE SQUARE ROOT.

The product of a number multiplied by itself, is called a square; and for this reason, the number, considered in relation to such a product, is called a SQUARE ROOT. For example, when we multiply 12 by 12, the product, 144, is a square, and 12 is the square root of 144.

If the root contains a decimal, the square will also contain a decimal of double the number of places; for example, 2.25 is the square of 1.5; and vice versâ, if the square contains a decimal, the square root will contain a decimal of half the number of places; for instance, 1.5 is the square root of 2.25.

In the upper line of the following table are arranged several square roots, and in the lower line, their squares.

Square roots.	1	2	3	4	5	6	7	8	9	10	11	12
Squares.	1	4	9	16	25	36	49	64	81	100	121	144

When the square of a mixed number is required, it may be reduced to an improper fraction, then squared, and reduced back to a mixed number.

The squares of the numbers from 3 to 5, increasing by $\frac{1}{4}$, are as follows.

Square roots.	3	$3\frac{1}{4}$	$3\frac{1}{2}$	$3\frac{3}{4}$	4	$4\frac{1}{4}$	$4\frac{1}{2}$	$4\frac{3}{4}$
Squares.	9	$10\frac{9}{16}$	$12\frac{1}{4}$	$14\frac{1}{16}$	16	$18\frac{1}{16}$	$20\frac{1}{4}$	$22\frac{9}{16}$

From this table we infer, that if a square contains a fraction, its square root also contains one; and vice versâ.

It is not possible to extract the square root of any number, which is not a perfect square; we can approximate the square root of such numbers, however, by the aid of decimals.

When a root is composed of two or more factors, we may multiply the squares of the several factors together, and the product will be the square of the whole root; and conversely, if a square be composed of two or more factors, each of which is a square, we have only to multiply together the roots of those factors, to obtain the complete root of the whole square. For example, 2304 $= 4 \times 16 \times 36$; the square roots of the factors are 2, 4, and 6; and $2 \times 4 \times 6 = 48$; and 48 is the square root of 2304, because $48 \times 48 = 2304$.

A square number cannot have more places of figures than double the places of the root; and, at least, but one less than double the places of the root. Take, for instance, a number consisting of any number of places, that shall be the greatest possible, of those places, as 99, the square of which is 9801, double the places of the root. Again, take a number consisting of any number of places, but let it be the least possible, of those places, as 10, the square of which is 100, one less than double the places of the root.

As the places of figures in the square cannot be more than double the number of places in the root, whenever we would extract the square root of any number, we point it off into periods of two figures each, by placing a dot over the place of units, another over hundreds, &c. Thus 1936. The places in the root can never be more or less in number, than the number of periods thus pointed off. When the number of places in the given sum is an odd number, the left hand period will contain only one figure, as 169; but the root will nevertheless consist of as many places as there are periods; for 13 is the square root of 169.

The terms, *square* and *square root*, are derived from geometry, which teaches us that the area of a square is found by multiplying one of its sides by itself.

14*

The word AREA signifies the quantity of space contained in any geometrical figure.

A SQUARE is a figure having four equal sides, and all its angles *right angles*. If we suppose the length of a side of the annexed square to be four feet, it is evident that the figure contains 4 times 4 small squares, each of which is 1 foot in length and 1 foot in breadth; and since a foot in length and a foot in breadth 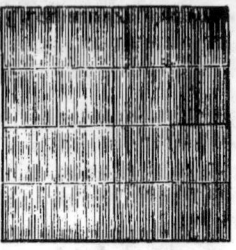 constitute a square foot, the whole square contains 4 times 4, or 16 square feet. If, instead of 4 feet, the length of a side were 4 yards, the whole square would contain 16 square yards; &c. Hence it is evident that the area, which is 16, is found by multiplying a side of the square by itself.

A PARALLELOGRAM is an oblong figure, having two of its sides equal and parallel to each other, but not of the same length with the other two, which are also equal and parallel to each other. We find the area, or contents of a parallelogram by multiplying the length by the breadth. If we suppose the annexed right angled parallelogram to be 8 feet long and 2 feet wide, it is manifest that it contains 2 times 8, or 16 square feet; if the length were 8 yards and the breadth 2 yards, it would contain 16 square yards; if 8 miles long and 2 miles wide, 16 square miles; &c. We see that the area of this parallelogram is the same with that of the preceding square; therefore the square root of the area of a parallelogram gives the side of a square equal in area to the parallelogram.

It is further to be observed, that the square root of the area of any geometrical figure whatever, is the side of a square, equal in area to the figure.

When the area of a square is given, the process of finding one of its sides, which is the root, is called the extraction of the square root, the principles of which we will now proceed to explain.

We have already learned, that a square number is a product resulting from two equal factors. For example, 2025 is a square number resulting from the multiplication of 45 by 45. To investigate the constituent parts of this product, we will separate the root into two terms, thus, $40 + 5$, and multiply it by itself in this form. We begin with multiplying $40 + 5$ by 5, and set down the products separately, which are $200 + 25$; we then multiply $40 + 5$ by 40, and set down the products separately, which are $1600 + 200$; the whole product, therefore, is $1600 + 200 + 200 + 25 = 2025$; thus we see, that the whole product or square contains the square of the first term, $40 \times 40 = 1600$; twice the product of the two terms, $40 \times 5 \times 2 = 400$; and the square of the last term, $5 \times 5 = 25$.

Now the extraction of the square root is the reverse of squaring or raising to the second power; therefore, the operation of extracting the square root of 2025, which we know is the square of 45, must be performed in the inverted order of raising 45 to the second power.

We will now extract the square root of 2025, and explain the process, step by step.

$$2025 \ (45$$
$$16$$

Divisor. $40 \times 2 = 80$ 425 dividend
Divisor, increased by last fig. 85   425 product of 85 by 5.

Explanation of the process. We began by separating the given number into periods of two figures each, putting a dot over the place of units, and another over hundreds, and thereby ascertained that the root would contain two places of figures. We then found that the greatest square in the left hand period was 16, and placed its root, which is 4, in the quotient, and subtracted the square from the left hand period, and to the remainder brought down the next period for a dividend.

Then, knowing the figure in the root to be in the place of tens, and therefore equal to 40, and that the second figure in the root must be such, that twice the product of the first and second terms, together with the square of the second, would complete the square, we took twice the root already found, viz. $40 \times 2 = 80$, for one of the factors, and using it for a divisor, found the second figure in the root by dividing the dividend by this factor.

Lastly, after finding that the second figure in the root was 5, we added it to the divisor, making $80 + 5 = 85$, and multiplied the sum by 5, the last figure in the root, and thus obtained twice the product of the two terms, and the square of the last term; because, 80 is twice the first term of the root, and being multiplied by 5, which is the last term, the result is twice the product of the two terms; and 5 being multiplied by 5, the product is the square of the last term.

It will be observed, that 4, the first figure in the root, being in the place of tens, was called 40, and doubled for a divisor; but, if we had merely doubled the root without any regard to its place, making the divisor 8, and had cut off the right hand figure of the dividend and divided what was left, the result would have been the same; because, in this operation, both divisor and dividend would have been divided by 10. Thus 8 is contained 5 times in 42. The figure obtained for the root, in this abridged method, would be placed at the right hand of the divisor, instead of being added thereto; thus, 85, making the completed divisor the same as before. This course, being the most concise, will be adopted in the rule for extracting the square root, which we shall hereafter give.

Suppose 169 square rods of land are to be laid out in a square, and the length of one of its sides is required.

We know that the length of a side must be such a number of rods, as, when multiplied by itself, will produce 169; therefore we must extract the square root of the given number of square rods, and that root will be the answer. $\sqrt{169} = 13$ *Ans.*

We here illustrate this
last example geometrical-
ly, by a square figure A B
C D, each side of which
is 13 rods long. This
square is divided into 169
small squares, each of
which is a square rod.

The whole figure is also
subdivided into four fig-
ures, two of which, e f g
D and h f i B, are squares;
and the other two, A h f e
and C i f g, are oblongs.

The square e f g D is 10 rods on a side, and, therefore,
contains 100 square rods. The oblongs are each 10 rods
long and 3 rods wide, and consequently each contains 30
square rods. The other square, B h f i, is 3 rods on a
side, and contains 9 square rods.

To illustrate the process of extracting the square root,
we shall take the side A B, which is divided into two
parts, the first of which, A h, is 10 rods long; the other,
h B, 3 rods long. A h being equal to e f, the square
of A h, the first part, gives the area of the square D e f g;
h B being equal to h f, the area of the oblong A h f e, is
found by multiplying the two parts, A h and h B, together; the area of the other oblong i f g C, is the same; therefore, the area of the two oblongs is twice the product
of the two parts, A h and h B. The square of the last
part, h B, gives the area of the square h B i f.

We have therefore the square of the first part A h,
$10 \times 10 = 100$ rods; twice the product of the two parts,
A h and h B, $10 \times 3 \times 2 = 60$ rods; and the square of
the last part h B, $3 \times 3 = 9$ rods. These being added
together make 169 rods, the square of the whole figure
A B C D.

This illustration of a square corresponds exactly with
that of the first example, and of course the extraction of
the square root must proceed on the principles there
exhibited.

From the illustrations of the two preceding examples, we give the following rule for the extraction of the square root.

RULE.   First—*Point off the given number into periods of two figures each, by putting a dot over the place of units, and another over every second figure to the left; and also to the right, when there are decimals.*

Secondly—*Find the greatest square in the left hand period, and write its root in the quotient.   Subtract the square of this root from the left hand period, and to the remainder bring down the next period for a dividend.*

Thirdly—*Double the root already found, for a divisor. Ascertain how many times the divisor is contained in the dividend, excepting the right hand figure, and place the result in the root, and also at the right hand of the divisor.   Multiply the divisor, thus increased, by the last figure in the root, and subtract the product from the dividend, and to the remainder bring down the next period for a new dividend.*

Fourthly—*Double the root already found for a new divisor, and continue to operate as before, until all the periods are brought down.*

It will sometimes happen, that, by dividing the dividend as directed in the rule, the figure, obtained for the root, will be too great.   When this happens, take a less figure, and go through the operation again.

When the places in the decimal are not an even number, they must be made so, by continuing the decimal, if it can be continued; if it cannot, by annexing a cipher, that the periods may be full.

If there be a remainder after all the periods are used, a period of decimal ciphers may be added; or, if the given number end in a decimal, the two figures that would arise from a continuation of the decimal.   The operation may be thus continued to any degree of exactness.

If any dividend shall be found too small to contain the divisor, put a cipher in the root, and bring down the next period to the right hand of the dividend for a new dividend, and proceed in the work.

When the square root of a mixed number is required, it will sometimes be necessary to reduce it to an improper fraction, or the vulgar fraction to a decimal, before extracting the root.

If either the numerator or denominator of a vulgar fraction be not a square number, the fraction must be reduced to a decimal, and the approximate root extracted.

1. Extract the square root of 4579600.

$$4579600(2140 \text{ Ans.}$$
$$4$$

1st. divisor	4	57 first dividend.
	41	41
2d. divisor	42	1696 second dividend.
	424	1696
		00

2. What is the square root of $110\frac{24}{99}$?

$$110.24(10.499 + \text{ Ans.}$$
$$1$$

1st. divisor,	2	10 first dividend.
2d. divisor,	20	1024 second dividend.
	204	816
3d. divisor,	208	20324 third dividend.
	2089	18801
4th. divisor,	2098	202324 fourth dividend.
	20989	198901
		13423 remainder.

Reducing $\frac{24}{99}$ to a decimal, we found it to be infinite, in the recurrence of 24 continually; therefore, in continuing the extraction of the root, instead of adding periods of decimal ciphers, we added the period 24 each time. The extraction of the root might have been continued indefinitely; but having obtained five places of figures in the root, we stopped, and marked off the three last places of the root for decimals; because we made use of three periods of decimals in the question.

3. What is the square root of 2704?
4. Extract the square root of 361.
5. What is the square root of 3025?
6. What is the square root of 121?
7. Extract the square root of 289.
8. Extract the square root of 400.
9. What is the square root of 4761?
10. What is the square root of 848241?
11. Extract the square root of 3356224.
12. What is the square root of 824464?
13. Find the square root of 49084036.
14. What is the square root of 688900?
15. Find the square root of 82864609.
16. Find the square root of 3684975616.
17. What is the square root of 44890000?
18. What is the square root of 165649?
19. Find the square root of 90484249636.
20. Find the square root of 264946225227849.
21. Find the square root of 262400.0625.
22. What is the square root of 841806.25?
23. What is the square root of 39.037504?
24. Find the square root of 213.715161.
25. Find the square root of .66650896.
26. What is the square root of $133407\frac{9}{16}$?
27. What is the square root of $15\frac{1}{64}$?
28. Extract the square root of $318\frac{1}{36}$.
29. What is the square root of $51\frac{1}{49}$?
30. Extract the square root of $2\frac{7}{9}$.
31. What is the square root of $556\frac{24}{25}$?
32. Extract the square root of $1096\frac{28}{81}$.
33. Find the square root of 4120900.
34. Extract the square root of 5.
35. Extract the square root of 8.
36. Extract the square root of 84.
37. Extract the square root of 99.
38. Extract the square root of 101.
39. Extract the square root of 120.
40. Extract the square root of 124.
41. Extract the square root of 143.
42. Extract the square root of 1.5.

43. Extract the square root of .00032754.
44. Extract the square root of 2.3.
45. Extract the square root of $\frac{2}{3}$.
46. Extract the square root of $\frac{3}{4}$.
47. Extract the square root of $\frac{2}{5}$.
48. Extract the square root of $\frac{22}{36}$.
49. Extract the square root of $113\frac{2}{5}$.
50. Extract the square root of $267\frac{3}{4}$.

The square root of the product of any two numbers is a mean proportional between those numbers.

Thus, 4 is a mean proportional between 2 and 8; because $2 : 4 = 4 : 8$. But when four numbers are proportionals, the product of the extremes is equal to the product of the means; that is, the product of the two given numbers is equal to the square of the mean proportional.

51. Find a mean proportional between 4 and 256.
52. Find a mean proportional between 4 and 196.
53. Find a mean proportional between 2 and 12.5.
54. Find a mean proportional between 9.8 and 5.
55. Find a mean proportional between 25 and 121.
56. Find a mean proportional between 180.625 and 10.
57. Find a mean proportional between 52 and $54\frac{1}{12}$.
58. Find a mean proportional between $\frac{7}{8}$ and $3\frac{1}{2}$.
59. Find a mean proportional between 12 and 147.
60. Find a mean proportional between $\frac{1}{4}$ and 4.
61. Find a mean proportional between .5 and 98.
62. Find a mean proportional between $4062\frac{55}{207}$ and 828.
63. Find a mean proportional between .25 and 1.
64. Find a mean proportional between .1 and 810.
65. Find a mean proportional between .04 and .36.
66. Find a mean proportional between .09 and .49.
67. Find a mean proportional between .2 and .018.

When the square root of the product of the two given numbers cannot be extracted without a remainder, the mean proportional is a SURD, and may be approximated by the aid of decimals.

68. Find a mean proportional between 6 and 12.

15

69. Find a mean proportional between 25 and 14.

70. Find a mean proportional between 64 and 21.

71. Find a mean proportional between 46 and 55.

72. Find a mean proportional between 5 and 81.

73. Find a mean proportional between 77 and 19.

74. A number of men spent 1 pound 7 shillings in company, which was just as many pence for each man, as there were men in the company. How many were there?

75. A company of men made a contribution for a charitable purpose; each man gave as many cents, as there were men in the company. The sum collected was 31 dollars 36 cents. How many men did the company consist of?

76. If you would plant 729 trees in a square, how many rows must you have, and how many trees in a row?

77. A certain regiment consists of 625 men. How many must be placed in rank and file, to form the regiment into a square?

78. It is required to lay out 40 acres of land in a square. Of what length must a side of the square be?

79. It is required to lay out 20 acres of land in the form of a right angled parallelogram, which shall be twice as long as it is wide. What will be its length and breadth? (See page 162.)

80. It is required to lay out 30 acres of land in the form of a right angled parallelogram, the length of which shall be three times the width. How long and how wide will it be?

A TRIANGLE is a figure having three sides and three angles. When one of the angles is such as would form one corner of a square, the figure is called a *right-angled* triangle, and the following propositions belong to it.

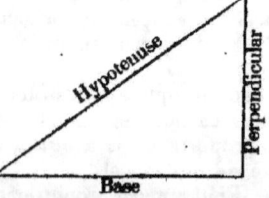

PROPOSITION 1st. *The square of the hypotenuse is equal to the sum of the squares of the other two sides.*

PROPOSITION 2d. *The square root of the sum of the squares of the base and perpendicular is equal to the hypotenuse.*

PROPOSITION 3d. *The square root of the difference of the squares of the hypotenuse and base is equal to the perpendicular.*

PROPOSITION 4th. *The square root of the difference of the squares of the hypotenuse and perpendicular is equal to the base.*

By observing the above propositions, when any two sides of a right-angled triangle are given, we may always find the remaining side. For example, suppose the base of the preceding figure to be 4 yards in length, and the perpendicular to be 3 yards in height; then the square of the base is 16 yards, and the square of the perpendicular 9 yards, and the sum of their squares is 25 yards. The square root of 25 yards is 5 yards, which is the length of the hypotenuse.

81. A certain castle, which is 45 feet high, is surrounded by a ditch, 60 feet broad. What must be the length of a ladder, to reach from the outside of the ditch to the top of the castle ?

82. A ladder 40 feet long, resting on the ground at the distance of 24 feet from the bottom of a straight tree, and leaning against the tree, just reaches to the first limb. What is the length of the tree's trunk ?

83. Two brothers left their father's house, and went, one, 64 miles due west, the other, 48 miles due north, and purchased farms, on which they now live. How far from each other do they reside ?

84. James and George, flying a kite, were desirous of knowing how high it was. After some consideration, they perceived, that their knowledge of the square root, and of the properties of a right angled triangle, would enable them to ascertain the height. James held the line close to the ground, and George ran forward till he came directly under the kite; then measuring the distance from James to George, they found it to be 312 feet; and pulling in the kite, they found the length of line out, to be 520 feet. How high was the kite ?

85. A ladder, 40 feet long, was so placed in a street, as to reach a window 33 feet from the ground on one side, and when turned to the other side without changing the place of its foot, reached a window 21 feet high. The breadth of the street is required.

86. The distance between the lower ends of two equal rafters, in the different sides of a roof, is 32 feet, and the height of the ridge above the foot of the rafters is 12 feet. Find the length of a rafter.

A straight line, drawn through the centre of a square, or through the centre of a right-angled parallelogram, from one angle to its opposite, is called a DIAGONAL; and this diagonal is the hypotenuse of both the right-angled triangles into which the square or parallelogram is thus divided.

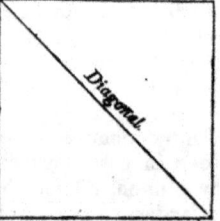

87. A certain lot of land, lying in a square, contains 100 acres: at what distance from each other are the opposite corners?

88. There is a square field containing 10 acres: what is the distance of the centre from either corner?

A CIRCLE is a plane surface bounded by one curve line, called the *circumference,* every part of which is equally distant from the centre.

A straight line through the centre of a circle is called a *diameter,* and a straight line from the centre of a circle to the circumference is called a *radius.*

The areas of all circles are to one another, as the squares of their like dimensions. That is, the area of a greater circle is to the area of a less circle, as the square of the diameter of the greater to the square of the diameter of the less. Or thus, the area of the greater is to the area of the less, as the square of the circumference of the greater to the square of the circumference of the less.

Therefore, to find a circle, which shall contain 2, 3, 4, &c. times more or less space than a given circle, we have the following—

RULE. *Square one of the dimensions of the given circle, and, if the required circle be greater, multiply the square by the given ratio, then the square root of the product will be the like dimension of the required circle; but, if the required circle be less than the given one, divide the square by the given ratio; then the square root of the quotient will be the similar dimension of the circle required.*

89. The diameter of a given circle is 11 inches: what is the diameter of a circle containing 9 times as much space?

90. Find the diameter of a circle, which shall contain one fourth of the area of a circle of 42 feet diameter.

91. What must be the circumference of a circular pond, to contain 4 times as much surface, as a pond, of $1\frac{1}{2}$ mile in circumference?

92. Find the circumference of a pond which shall contain $\frac{1}{16}$ part as much surface, as a pond of $13\frac{1}{2}$ miles circumference.

93. Find the diameter of a circle, which shall be 36 times as much in area, as a circle of $18\frac{3}{4}$ rods diameter.

The diameter of a circle is to the circumference in the ratio of 1 to 3.14159265, nearly: therefore, if we know the one, we can find the other. Thus, the circumference of a circle, the diameter of which is 8, is 3.14159265 × 8 = 25.1327412; the diameter of a circle, the circumference of which is 15.70796325, is 15.70796325 ÷ 3.14159265 = 5.

To find the area of a cirlce, multiply the circumference by the radius, and divide the product by 2.

94. How many feet in length is the side of a square, equal in area to a circle of 36 feet diameter?

95. Find the side of a square equal in area to a circle of 20 rods in diameter.

96. Find the diameter of a pond, that shall contain $\frac{1}{4}$ as much surface, as a pond of 6,986 miles circumference.

97. Find the length and breadth of a right-angled parallelogram, which shall be 4 times as long as it is wide, and equal in area to a circle of 43.9822971 rods circumference.

98. Find the circumference of a pond, which shall contain as much surface, as 9 ponds of $\frac{1}{4}$ of a mile diameter each.

XXX.

EXTRACTION OF THE CUBE ROOT.

A CUBE is represented by a solid block—like either of those annexed—with six plane surfaces; having its length, breadth, and height all equal. Consequently, the solid contents of a cube are found by multiplying one of its sides twice into itself. For this reason, the third power of any number is called a cube.

Therefore, if we multiply the square of a number by its root, we obtain a product, which is called a cube, or a cubic number. For instance, 4 multiplied by 4 produces 16, which is the square of 4, as shown on one of the sides of this larger block; and 16 multiplied by 4 produces 64, which is the cube of 4, as shown by the whole of the larger block.

Thus the cube of any quantity is produced by multiplying the quantity by itself, and again multiplying the product by the original quantity. When the quantity to be

cubed is a mixed number, it may be reduced to an improper fraction, and the fraction cubed, and then reduced
back to a mixed number.

As we can, in the manner explained, find the cube of
a given number, so also, when a number is proposed, we
may reciprocally find a number, which being cubed will
produce the given number. In this case, the number
sought is called, in relation to the given number, the
CUBE ROOT. Therefore, the cube root of a given number is the number, whose cube is equal to the given
number. For instance, the cube root of 125 is 5; the
cube root of 216 is 6; the cube root of $\frac{1}{8}$ is $\frac{1}{2}$; the cube
root of $3\frac{3}{8}$ is $1\frac{1}{2}$.

A cube cannot have more places of figures than triple
the places of the root, and, at least, but two less than
triple the places of the root. Take, for instance, a
number consisting of any number of places, that shall be
the greatest possible in those places, as 99, the cube of
which is 970299; here the places are triple. Again, take
a number, that shall be the least possible in those places,
as 10, the cube of which is 1000; here the places are two
less than triple.

It is manifest from what has been said, that a cubic
number is a product resulting from three equal factors.
For example, 3375 is a cubic number arising from $15 \times$
15×15. To investigate the constituent parts of this cubic
number, we will separate the root, from which it was
produced, into two parts, and instead of 15, write $10 + 5$,
and raise it to the third power in this form:

$$
\begin{array}{r}
10 + 5 \\
10 + 5 \\
\end{array}
$$

Product of $10 + 5$ by 5, - - - -	$50 + 25$
Product of $10 + 5$ by 10, - - -	$100 + 50$
The square, - - - - -	$100 + 100 + 25$
	$10 + 5$
Prod. of $100 + 100 + 25$ by 5, - -	$500 + 500 + 125$
Prod. of $100 + 100 + 25$ by 10,	$1000 + 1000 + 250$
The third power, - -	$1000 + 1500 + 750 + 125$

This product contains the cube of the first term, three

times the square of the first term multiplied by the second term, three times the first term multiplied by the square of the second term, and the cube of the second term: thus, $10 \times 10 \times 10 = 1000$; $10 \times 10 \times 3 \times 5 = 1500$; $10 \times 3 \times 25 = 750$; $5 \times 5 \times 5 = 125$.

Now, if the cube be given, viz. $1000 + 1500 + 750 + 125$, and we are required to find its root, we readily perceive by the first term 1000, what must be the first term of the root, since the cube root of 1000 is 10; if, therefore, we subtract the cube of 10, which is 1000, from the given cube, we shall have for a remainder, $10 \times 10 \times 3 \times 5 = 1500$, $10 \times 3 \times 25 = 750$, and $5 \times 5 \times 5 = 125$; and from this remainder we must obtain the second term in the root. As we already know that the second term is 5, we have only to discover how it may be derived from the above remainder. Now that remainder may be expressed by two factors; thus, $(\overline{10 \times 10 \times 3} + \overline{10 \times 3 \times 5} + 5 \times 5)\, 5$: therefore, if we divide by three times the square of the first term of the root, plus three times the first term multiplied by the second term, plus the square of the second term, the quotient will be the second term of the root, which is 5.

But, as the second term of the root is supposed to be unknown, the divisor also is unknown; nevertheless we have the first term of the divisor, viz. three times the square of the root already found; and by means of this, we can find the next term of the root, and then complete the divisor, before we perform the division. After finding the second term of the root, it will be necessary, in order to complete the divisor, to add thrice the product of the two terms of the root, and the square of the second term, to three times the square of the first term previously found.

The preceding analysis explains the following rule for the extraction of the cube root.

RULE. First—*Point off the given number into periods of three figures each, beginning at the unit's place, and pointing to the left in integers, and to the right in decimals; making full periods of decimals by supplying the deficiency, when any exists.*

2dly—*Find the root of the left hand period, place it in the quotient, and subtract its cube from the given number. The remainder is a new dividend.*

3dly—*Square the root already found and multiply its square by 3, for a divisor.*

4thly—*Find how many times the divisor is contained in the dividend, and place the result in the quotient.*

5thly—*In order to complete the divisor, multiply the root previously found, by the number last put in the root, triple the product, and add the result to the divisor; also square the number last put in the root, and add its square to the divisor.*

Lastly—*Multiply the divisor thus completed, by the number last put in the root, and subtract the product from the dividend. The remainder will be a new dividend.*

Thus proceed, till the whole root is extracted.

We will extract the cube root of 34965783, denoting each step of the operation, from *first* to *last*, by a reference to that part of the rule, under which it falls.

First, - - - - - - - - -			34965783
2dly.	Cube of 300, subtracted, - -		27000000(300
	New dividend, - - - - -		7965783
3dly.	300×300×3 [a divisor]	270000	
4thly.	Divisor in new dividend, - -		- - - (20
5thly.	Triple prod. of 300×20,	18000	
	Square of last number,	400	
	Divisor completed, -	288400	
Lastly.	288400 × 20, subt'ed, - -		5768000
	New dividend, - - - -		2197783
3dly.	320×320×3 [a divisor]	307200	
4thly.	Divisor in new dividend, - -		- - - (7
5thly.	Triple prod. of 320×7,	6720	
	Square of last number,	49	
	Divisor completed, -	313969	
Lastly.	313969 × 7, subtracted,		2197783

$$300 + 20 + 7 = 327 \quad \textit{Ans.}$$

In completing every divisor, we have three products to add together; viz. three times the square of the root already found; three times the product resulting from the multiplication of the root already found, by the number last put in the root; and the square of the last number.

If the ciphers be removed from the right hand of each of these products, the remaining figures in each succeeding product will stand one place to the right of each preceding product; therefore, the work will be considerably abridged by adopting the following—

RULE. First—*Point off the given number into periods of three figures each, as before directed.*

2dly—*Find the root of the left hand period, place it in the quotient without regard to local value, and subtract its cube from that period; and to the remainder bring down the next period for a dividend.*

3dly—*Square the root already found, without any regard to its local value, and multiply its square by 3, for a divisor.*

4thly—*Find how many times the divisor is contained in the dividend, omitting the two right hand figures, and place the result in the quotient.*

5thy—*To complete the divisor, multiply the root previously found, by the figure last placed in the quotient, without regarding local value, triple the product, and write it under the divisor, one place to the right; square the figure last put in the quotient, and write its square under the preceding product, one place to the right. Add these three together, and their sum is the divisor completed.*

Lastly—*Multiply the divisor thus completed, by the figure last placed in the root, and subtract the product from the dividend; and to the remainder bring down the next period for a new dividend.*

Thus proceed, till the whole root is extracted.

Observe, that, when the divisor is not contained in the dividend, as sought in the fourth part of the rule, a cipher must be put in the root, and the next period brought down for a new dividend.

Observe, also, that when the figure obtained for the root by dividing, as directed in the fourth part of the rule, is found, on completing the divisor, to be too large, a smaller figure must be substituted in its place, and the divisor completed anew.

There are always as many decimals in the root, as periods of decimals in the power.

We will extract the cube root of 65890311319, in the abridged form; referring, as before, to the particular part of the rule, under which each step of the operation proceeds.

First,			65890311319(4039
2dly.	Cube of 4, subt'd - - -		64
	Dividend, - - - - -		1890
3dly.	4×4×3 [div'r] - -	48	
4thly.	48 was not con-		
	tained in 18. - - -		0000
Lastly.	New dividend, - - -		1890311
3dly.	40×40×3, - -	4800	
4thly.	4800 in 18903,		
	3 times.		
5thly.	Triple product		
	of 40×3, - - -	360	
	Square of 3, - - -	9	
	Divisor comp'd, -	483609	
Lastly.	483609×3, and		
	subtracted, - - - -		1450827
	New dividend, - - -		439484319
3dly.	403×403×3, -	487227	
4thly.	487227 in 4394		
	843, 9 times.		
5thly.	Triple product		
	of 403×9, -	10881	
	Square of 9 -	81	
	Divisor comp'd,	48831591	
Lastly.	48831591 × 9,		
	and subtracted, - - -		439484319

Ans. 4039

We will now extract the cube root of 178263.433152,
in the abridged form, as in the preceding example; but
without reference to the parts of the rule.

```
                178263.433152(56.28   Ans.
                125

        75 |   53263
        90
        36
      8436 |   50616
      9408 |   2647433
       336
         4
    944164 |   1888328
    947532 |   759105152
     13488
        64
  94888144 |   759105152
                . . . . . . . . .
```

1. Extract the cube root of 614125.
2. What is the cube root of 191102976 ?
3. What is the cube root of 18399.744 ?
4. Find the cube root of 253395799552.
5. What is the cube root of 1740992427 ?
6. Extract the cube root of 35655654571.
7. Find the cube root of 27243729729.
8. What is the cube root of 912673000000 ?
9. What is the cube root of 67518581248 ?
10. Find the cube root of 729170113230343.
11. Extract the cube root of 643.853447875.
12. Find the cube root of .000000148877.
13. What is the cube root of 123 ?
14. Extract the cube root of 517.
15. Extract the cube root of 900.
16. Extract the cube root of $\frac{4}{15}$.
17. What is the cube root of $\frac{8}{9}$?
18. What is the cube root of $\frac{8}{27}$?
19. What is the cube root of $\frac{12167}{148877}$?
20. Extract the cube root of 26.

To find two MEAN PROPORTIONALS between two given numbers, *divide the greater by the less, and extract the cube root of the quotient: then multiply the cube root by the least of the given numbers, and the product will be the least of the mean proportionals; and the least mean proportional multiplied by the same root, will give the greatest mean proportional.*

21. What are the two mean proportionals between 6 and 750?

22. What are the two mean proportionals between 56 and 12096?

To find the side of a cube equal in solidity to any given solid, *extract the cube root of the solid contents of the given body, and it will be the required side.*

23. There is a stone, of cubic form, containing 21952 solid feet. What is the length of one of its sides?

24. The solid contents of a globe are 15625 cubic inches: required the side of a cube of equal solidity.

25. Required the side of a cubical pile of wood, equal to a pile 28 feet long, 18 ft. broad, and 4 ft. high.

All solid bodies are to each other, as the cubes of their diameters, or similar sides.

26. If a ball 6 inches in diameter weighs 32 pounds, what is the diameter of another ball of the same metal, weighing 4 pounds?

27. If a ball of 4 inches diameter weighs 9 pounds, what is the diameter of a ball weighing 72 pounds?

28. What must the side of a cubic pile of wood measure, to contain $\frac{1}{8}$ part as much as another cubic pile, which measures 10 feet on a side?

29. If 8 cubic piles of wood, each measuring 8 feet on a side, were all put into one cubic pile, what would be the dimensions of one of its sides?

30. The solid contents of a globe 21 inches in diameter are 4849.0596 solid inches; what is the diameter of a globe, whose solid contents are 11494.0672 inches?

31. What are the inside dimensions of a cubical bin, that will hold 85 bushels of grain? (See note, page 27.)

32. What must be the inside dimensions of a cubical

bin, to hold 450 bushels of potatoes, 2815.489 cubic inches, (heaped measure), making a bushel?

33. What must be the inside measure of a cubical cistern, to hold 10 hogsheads of water?

34. What must be the inside measure of a cubical cistern, that will hold 20 hogsheads of water?

35. What are the inside dimensions of a cubical cistern, that holds 40 hogsheads of water?

36. Suppose a chest, whose length is 4 feet 7 inches, breadth 2 feet 3 inches, and depth 1 foot 9 inches: what is the side of a cube of equal capacity?

37. Suppose I would make a cubical bin of sufficient capacity to contain 108 bushels; what must be the dimensions of the sides?

XXXI.

ROOTS OF ALL POWERS.

The roots of many of the higher powers may be extracted by repeated extractions of the square root, or cube root, or both, as the given power may require. Whenever the index of the given power can be resolved into factors, these factors denote the roots, which, being successively extracted, will give the required root.

Thus, the index of the fourth power is 4, the factors of which are 2×2; therefore, extract the square root of the fourth power, and then the square root of that square root will be the fourth root. The sixth root is the cube root of the square root, or the square root of the cube root; because $3 \times 2 = 6$. The eighth root is the square root of the square root of the square root; because $2 \times 2 \times 2 = 8$. The ninth root is the cube root of the cube root; because $3 \times 3 = 9$. The tenth root is the fifth root of the square root; because $2 \times 5 = 10$. The twelfth root is the cube root of the square root of the square root; because $2 \times 2 \times 3 = 12$. The twenty-seventh root is the cube root of the cube root of the cube root; because $3 \times 3 \times 3 = 27$.

The following is a GENERAL RULE for extracting the roots of all powers.

RULE. First—*Prepare the given number for extraction, by pointing off from the unit's place, as the required root directs; that is, for the fourth root, into periods of four figures; for the fifth root into periods of five figures, &c.*

2dly.—*Find the first figure of the root by trial, and subtract its power from the left hand period.*

3dly.— *To the remainder bring down the first figure in the next period for a dividend.*

4thly.—*Involve the root to the next inferior power to that which is given, and multiply it by the number denoting the given power, for a divisor.*

5thly.—*Find how many times the divisor is contained in the dividend, and the quotient will be another figure of the root.*

6thly.—*Involve the whole root to the given power, and subtract it from the two left hand periods of the given number.*

Lastly.—*Bring down the first figure of the next period to the remainder, for a new dividend, to which find a new divisor, as before. Thus proceed, till the whole root is extracted.*

Observe, that when a figure obtained for the root by dividing, is found by involving, to be too great, a less figure must be taken, and the involution performed again.

We will extract the fifth root of 36936242722357.

$$36936242722357(517 \; Ans.$$

$$5^5 = \quad 3125$$

$5^4 \times 5 = 3125$ first divisor. $\underline{5686}$ first dividend.

$$51^5 = \quad 345025251$$

$51^4 \times 5 = 33826005,$ }
   second divisor. } 243371762 2d. dividend.

$$517^5 = \quad 36936242722357$$

1. What is the fifth root of 5584059449 ?
2. Find the fifth root of 2196527536224.
3. Extract the fifth root of 16850581551 ?
4. Find the seventh root of 2423162679857794647.

XXXII.

EQUIDIFFERENT SERIES.

A series of numbers composed of any number of terms, which uniformly increase or decrease by the same number, is called an EQUIDIFFERENT SERIES. This series has, very commonly, but without any propriety, been called *Arithmetical Progression*.

When the numbers increase, they form an ascending series; but when they decrease, a descending series. Thus, the natural numbers, 1, 2, 3, 4, 5, 6, 7, 8, 9, form an ascending series, because they continually increase by 1; but 9, 8, 7, 6, &c. form a descending series, because they continually decrease by 1.

The numbers, which form the series, are called the *terms* of the series. The first and last terms in the series are called the *extremes*; and the other terms, the *means*.

The number, by which the terms of the series are continually increased or diminished, is called the *common difference*. Therefore, when the first term and common difference are given, the series may be continued to any length. For instance, let 1 be the first term in an equidifferent series, and 3 the common difference, and we shall have the following increasing series; 1, 4, 7, 10, 13, &c., in which each succeeding term is found by adding the common difference to the preceding term.

THEOREM I. *When four numbers form an equidifferent series, the sum of the two extremes is equal to the sum of the two means.* Thus, 1, 3, 5, 7, is an equidifferent series, and $1+7=3+5$. Also in the series 11, 8, 5, 2, $11+2=8+5$.

THEOREM II. *In any equidifferent series, the sum of the two extremes is equal to the sum of any two means, that are equally distant from the extremes; and equal to double the middle term, when there is an uneven number of terms.* Take, for example, the equidifferent series, 2, 4, 6, 8, 10, 12, 14; $2+14=4+12$; and $2+14=6+10$; also $2+14=8+8$.

. Since, from the nature of an equidifferent series, the second term is just as much greater or less than the first, as the last but one is less or greater than the last, it is evident, that when these two means are added together, the excess of the one will make good the deficiency of the other, and their sum will be the same with that of the. two extremes.   In the same manner it appears, that the sum of any other two means equally distant from the extremes, must be equal to the sum of the extremes.

THEOREM III.   *The difference between the extreme terms of an equidifferent series is equal to the common difference multiplied by the number of terms less 1.* Thus, of the six terms, 2, 5, 8, 11, 14, 17, the common difference is 3, and the number of terms less 1, is 5; then the difference of the extremes is 17—2, and the common difference multiplied by the number of terms less 1, is 3×5; and $17 - 2 = 3 \times 5$.

The difference between the first and last terms, is the increase or diminution of the first by all the additions or subtractions, till it becomes equal to the last term: and, as the number of these equal additions or subtractions is one less than the number of terms, it is evident that this common difference being multiplied by the number of terms less 1, must give the difference of the extremes.

THEOREM IV.   *The sum of all the terms of any equidifferent series is equal to the sum of the two extremes multiplied by the number of terms and divided by 2; or, which amounts to the same, the sum of all the terms is equal to the sum of the extremes multiplied by half the the number of terms.*   For example, the sum of the following series, 2, 4, 6, 8, 10, 12, 14, 16, is $\overline{2 + 16} \times 4 = 72$. This is made evident by writing under the given series the same series inverted, and adding the corresponding terms together as follows.

The given series,	2, 4, 6, 8, 10, 12, 14, 16.
Same series inverted,	16, 14, 12, 10, 8, 6, 4, 2.
. Sums of the series,	18, 18, 18, 18, 18, 18, 18, 18.

This series of equal terms, (18), is evidently equal to twice the sum of the given series; but the sum of these equal terms is $18 \times 8 = 144$; and since this sum is twice as great as that of the given series, the sum of the given series must be 72.

Any three of the five following things being given, the other two may be readily found.

The *first* term.
The *last* term.
The *number* of terms.
The common *difference*.
The *sum* of all the terms.

PROBLEM. I.   The extremes and number of terms being given, to find the sum of all the terms.

RULE.   *Multiply the sum of the extremes by the number of the terms, and half the product will be the sum of all the terms.*   See Theorem 4th.

1. The first term in an equidifferent series, is 3, the last term 19, and the number of terms is 9.   What is the sum of the whole-series ?

2. How many strokes does a common clock strike in 12 hours ?

3. A hundred cents were placed in a right line, a yard apart, and the first a yard from a basket.   What distance did the boy travel, who, starting from the basket, picked them up singly, and returned with them one by one to the basket ?

4. If a number of dollars were laid in a straight line for the space of a mile, a yard distant from each other, and the first a yard from a chest, what distance would the man travel, who, starting from the chest, should pick them up singly, returning with them one by one to the chest ?

PROBLEM II.   The extremes and number of terms given, to find the common difference.

RULE.   *Subtract the less extreme from the greater, and divide the remainder by the number of terms less 1, and the quotient will be the common difference.*

It has been shown under Theorem 3d. that the difference of the extremes is found by multiplying the common difference by the number of the terms less 1; consequently, the common difference is found by dividing the difference of the extremes by the number of the terms less 1.

5. A man had 10 sons, whose ages differed alike; the youngest was 2 years old, and the eldest 29. What was the difference of their ages?

6. The extremes in an equidifferent series are 3 and 87, and the number of terms 43. Required the common difference.

7. A man is to travel from Boston to a certain place in 9 days, and to go but 5 miles the first day, and to increase his journey every day alike, so that the last day's journey may be 37 miles. Required the daily increase, and also the number of miles travelled.

PROBLEM III. The extremes and common difference given, to find the number of terms.

RULE. *Divide the difference of the extremes by the common difference, and add 1 to the quotient; the sum will be the number of terms.*

The difference of the extremes divided by the number of the terms less 1, gives the common difference; consequently, the same divided by the common difference must give the number of terms less 1: hence, this quotient augmented by 1, must give the number of terms.

8. The extremes in an equidifferent series are 3 and 39, and the common difference is 2: what is the number of terms?

9. A man going a journey, travelled 7 miles the first day, and increased his journey every day by 4 miles, and the last day's journey was 51 miles. How many days did he travel, and how far?

10. A man commenced a journey with great animation, and travelled 55 miles the first day; but on the second day he began to be weary, and travelled only 51 miles, and thus continued to lose 4 miles a day, till his last day's journey was only 15 miles. How many days did he travel?

PROBLEM IV.   To find an equidifferent mean between two given terms.

RULE.   *Add the two given terms together, and half their sum will be the equidifferent mean required.*

11. Find an equidifferent mean between 3 and 15.

12. What is the equidifferent mean between 7 and 53?

13. Find an equidifferent mean between 5 and 18.

PROBLEM V.   To find two equidifferent means between the given extremes.

RULE.   *Divide the difference of the extremes by 3, and the quotient will be the common difference, which, being continually added to the less extreme, or subtracted, from the greater, gives the two required means.*

14. Find two equidifferent means between 4 and 13.

15. Find two equidifferent means between 5 and 22.

16. Find two equidifferent means between 4 and 53.

PROBLEM VI.   To find any numbers of equidifferent means between the given extremes.

RULE.   *Divide the difference of the extremes by the required number of means plus 1, and the quotient will be the common difference, which being continually added to the less extreme, or subtracted from the greater, will give the mean terms required.*

17. Find five equidifferent means between 4 and 28.

18. Find six equidifferent means between 6 and 55.

19. Find 3 equidifferent means between 34 and 142.

20. Find one equidifferent mean between 56 and 100.

XXXIII.

CONTINUAL PROPORTIONALS.

The numbers of a series in which the successive terms increase by a common multiplier, or decrease by a common divisor, are CONTINUAL PROPORTIONALS.

This series of numbers has been commonly called a *Geometrical Progression;* but, perceiving no appropriate meaning in this term, we choose to call the series, what it is in truth, *a series of Continual Proportionals.*

The common multiplier, or common divisor, by which the successive terms are increased or deminished, is called the *ratio* of the series, or the *common ratio*.

Thus, $\overset{1}{3}, \overset{2}{6}, \overset{3}{12}, \overset{4}{24}, \overset{5}{48}, \overset{6}{96}$ is a series of continual proportionals, in which each successive term is produced by multiplying the preceding term by 2, which is the common ratio. The numbers 1, 2, 3, 4, &c. standing above the series, mark the place, which each term holds in the series.

Also, 729, 243, 81, 27, 9, 3, 1, is a series of continual proportionals, in which each successive term is found by dividing the preceding term by 3, the common ratio.

In an increasing series, the ratio is the quotient, which results from the division of the consequent by the antecedent; but in a decreasing series, the ratio is the quotient resulting from the division of the antecedent by the consequent.

In every series of continual proportionals, any four successive terms constitute a proportion. Thus, in the first of the above series, $3 : 6 = 12 : 24$, and $6 : 12 = 24 : 48$, also, $12 : 24 = 48 : 96$. In the second series, $729 : 243 = 81 : 27$, $243 : 81 = 27 : 9$, $81 : 27 = 9 : 3$, $27 : 9 = 3 : 1$. Therefore, when there are only four terms, the product of the extremes is equal to the product of the means.

Furthermore, in any series of continual proportionals, the product of the extremes is equal to the product of any two terms equally distant from them; and equal to the second power of the middle term, when there is an uneven number of terms. For instance, take the continual proportionals 2, 4, 8, 16, 32, 64, 128; then $2 \times 128 = 4 \times 64$; also $2 \times 128 = 8 \times 32$; and $2 \times 128 = 16 \times 16$.

When the first term and the ratio are given, a series of continual proportionals may be extended to any number of terms by continually multiplying by the ratio in an increasing series, or dividing in a decreasing series.

For example, the first term being 2, and the ratio 3, if we make it an increasing series by continually multiplying by the ratio, we obtain the following series, 2, 6, 18, 54, 162, 486, which may be extended to any number of terms; but, if we make it a decreasing series by continually dividing by the ratio, we obtain the following series, $2, \frac{2}{3}, \frac{2}{9}, \frac{2}{27}, \frac{2}{81}, \frac{2}{243}$, which may also be extended to any number of terms.

In the series 2, 6, 18, 54, 162, 486, we obtain the second term by multiplying the first term by the ratio; the third term by multiplying the second term by the ratio; the fourth term by multiplying the third term by the ratio; the fifth term by multiplying the fourth term by the ratio; the sixth term by multiplying the fifth term by the ratio. Therefore, since to obtain the sixth term, we have to multiply five times by the ratio, it is evident that we should also obtain the sixth term by multiplying the first term by the fifth power of the ratio. The fifth power of 3 is 243, which being multiplied into the first term, the result is 486, the same as in the series.

Hence we see, that any term in any increasing series of continual proportionals may be found by multiplying the first term by that power of the ratio, which is denoted by the number of terms preceding the required one. For instance, the ninth term in an increasing series is found by multiplying the first term by the eighth power of the ratio; thus let 2 be the first term, and 3 the common ratio; then 2×3^8 gives the ninth term, which is 13122.

If the series be a decreasing one, any term in it may be found by dividing the first term by that power of the ratio; which is denoted by the number of terms preceding the required one. For instance, the seventh term in a decreasing series is found by dividing the first term by the sixth power of the ratio; thus, let 24576 be the first term in a decreasing series, and 4 the common ratio; then $24576 \div 4^6$ gives the seventh term, which is 6.

We will now state several problems, which occur in continual proportionals, and give the rules for performing them.

PROBLEM I. The first term and the ratio being given, to find any other proposed term.

RULE. *Raise the ratio to a power, whose index is equal to the number of terms preceding the required term: then, if it be an increasing series, multiply the first term by this power of the ratio; but, if it be a decreasing series, divide the first term by it: the result will be the required term.*

1. Required the eighth term in an increasing series whose first term is 6, and ratio 2.

2. Required the ninth term in a decreasing series, the first term of which is 131072, and the ratio 4.

3. What is the seventh term in an increasing series, the first term being 3, and the ratio 1.5 ?

4. What is the sixth term in an increasing series, whose first term is $\frac{3}{16807}$, and ratio 7 ?

5. What is the tenth term in a decreasing series, the first term being 387420489, and the ratio 9 ?

One of the principal questions, which occurs in a series of continual proportionals, is to find the sum of the series. We shall, therefore, illustrate the method.

Let there be given the following series, consisting of seven terms, whose common ratio is 3; viz. 2, 6, 18, 54, 162, 486, 1458. Let each term in this series be multiplied by the ratio 3; and let each product be removed one place to the right of the terms in the given series.

The given series,   2, 6, 18, 54, 162, 486, 1458
multiplied by ratio.   6, 18, 54, 162, 486, 1458, 4374.

Now the last term in the second series is produced by multiplying the last term in the given series by the ratio; and it is evident that if the given series be subtracted from the second series, the remainder will be the last term in the second series diminished only by the first term in the given series, and this remainder will be twice the sum of the given series; consequently, if we divide it by 2, the quotient will be the sum of the given series; but 2 is the ratio less 1. Hence

PROBLEM II. The extremes and the ratio being given, to find the sum of the series.

RULE. *Multiply the greater extreme by the ratio, from the product subtract the less extreme, and divide the remainder by the ratio less 1, and the quotient will be the sum of the series.*

6. The first term in a series of continual proportionals is 1, the last term is 65611, and the ratio is 3. What is the sum of the series?

$$65611 \times 3 = 196833$$
$$1$$

ratio $3-1=2)\overline{196832}$

$$98416 \quad Ans.$$

7. The extremes of a series of continual proportionals are 3 and 12288, and the ratio is 4. What is the sum of the series?

8. The first term in a series of continual proportionals in 12500, the last term is 4, and the ratio 5. What is the sum of the series?

9. The first term in a series of continual proportionals is 7, the last term 1792, and the ratio is 2. What is the sum of the series?

10. The extremes in a series of continual proportionals are 5 and 37.96875, and the ratio is 1.5. What is the sum of the series?

11. The first term in a series of continual proportionals is 100, the last term .01, and the ratio 2.5. Required the sum of the series.

PROBLEM III. The first term, the ratio, and the number of terms given, to find the sum of the series.

RULE. *Find the last term by problem 1, and the sum of the series by problem 2.*

12. The first term in an increasing series of continual proportionals is 6, the ratio 4, and the number of terms 8. What is the sum of the series?

13. The first term in an increasing series of continual proportionals is $\frac{1}{5}$, the ratio 4, and the number of terms 13. Required the sum of the series.

14. The first term in a decreasing series of continual proportionals is 1, the ratio 3, and the number of terms 12. What is the sum of the series?

15. A man offers to sell his horse by the nails in his shoes, which are 32 in number. He demands one mill for the first nail, 2 for the second, 4 for the third, and so on, demanding for each nail twice the price of the preceding. It is required to find what would be the price of the horse.

16. An ignorant fop wanted to purchase an elegant house, and a facetious man told him he had one, which he would sell him on these moderate terms; viz. that he should give him one cent for the first door, 2 for the second, 4 for the third, and so on, doubling the price for every door, there being 36. It is a bargain, cried the simpleton, and here is a half-eagle to bind it. What was the price of the house ?

PROBLEM IV.   The extremes and the number of terms being given, to find the ratio.

RULE.   *Divide the greater extreme by the less, and the quotient will be that power of the ratio, which is denoted by the number of terms less* 1; *consequently, the corresponding root of this quotient will be the ratio.*

This problem is the reverse of problem 1, and the reasoning which precedes that problem, sufficiently elucidates the rule in this.

The first term in a series of proportionals is 192, the last term 3, and the number of terms 7. What is the ratio? $192 \div 3 = 64$; the number of terms less 1, is 6; therefore the sixth root of 64, which is 2, is the ratio.

17. In a series of continual proportionals, the first term is 7, the last 45927, and the number of terms 9. What is the ratio ?

18. The first term in a series of continual proportionals is 26244, the last term 4, and the number of terms 5. Required the ratio.

19. The first term in a series of continual proportionals is $\frac{1}{6}$, the last term $102942\frac{7}{8}$, and the number of terms 8. What is the ratio ?

20. The first term in a series of continual proportionals is 78125, the last term $\frac{1}{125}$, and the number of terms 11. Required the ratio.

17

PROBLEM V.   To find any number of mean proportionals between two given numbers.

RULE.   *The two given numbers are the extremes of a series consisting of two more terms than there are means required; hence the ratio will be found by problem 4. Then the product of the ratio and the less extreme will be one of the means; the product of this mean and the ratio will be another mean; and so on, till all the required means are found.*

When only one mean is required, it is the square root of the product of the extremes.

21. Find 3 mean proportionals between 5 and 1280.

Here the series is to consist of five terms, and the extremes are 5 and 1280; hence the ratio is found by the fourth problem to be 4; and by the repeated multiplication of the least term by the ratio, the means are found to be 20, 80, and 320.

22. Find four mean proportionals between ½ and 2401.

23. Find five mean proportionals between the numbers, 279936 and 6.

24. Find a mean proportional between 1 and 2809.

COMPOUND INTEREST BY SERIES.

It has been shown in ART. xv, page 107, that compound interest is that which arises from adding the interest to the principal at the end of each year, and taking the amount for a new principal.   Now, the several amounts for the several years form a series of continual proportionals; and, to find the amount for any number of years, we may adopt the following—

RULE.   *Find the last term of an increasing series of continual proportionals, whose first term is the principal, whose ratio is the amount of 1 dollar for 1 year, and whose number of terms is the number of years-plus 1.   The last term is the required amount.*   See Problem 1st.

In the examples under this rule, no more than six decimal places need be included.

25. What is the amount of $100, at 6 per cent. compound interest, for 4 years ?

26. What is the amount of $75, at 5 per cent. compound interest, for 9 years?

27. What is the amount of $294, at 4 per cent. compound interest, for 7 years?

28. Find the amount of $18.25, at 7 per cent. compound interest, for 12 years.

29. Find the amount of $751.30, at 5 per cent. compound interest, for 8 years.

30. Find the amount of $4798, at 6 per cent. compound interest, for 12 years.

31. What is the amount of $5.14, at 7 per cent. compound interest, for 16 years? What is the interest?

32. What is the compound interest of $1000 for 20 years, at 6 per cent.? What is the amount?

COMPOUND DISCOUNT.

Discount corresponding to simple interest has already been treated, in ART. XVI; but discount corresponding to compound interest is now to be computed.

On the supposition that money can be let out at compound interest, the *present worth* of a debt, payable at a future period without interest, is that principal, which, at *compound interest*, would give an *amount* equal to the debt, at the period when the debt is payable.

RULE. *Find the last term of a decreasing series of continual proportionals, whose first term is the debt, whose ratio is the amount of 1 dollar for 1 year, and whose number of terms is the number of years plus 1. The last term is the present worth.* See Problem 1st.

33. What principal, at 10 per cent. compound interest, will amount, in 4 years, to $8.7846?

34. What is the present worth of $68.40, payable 11 years hence; allowing discount according to 5 per cent. compound interest?

35. What is the present worth of $350, payable in 5 years; allowing discount at the rate of 6 per cent. compound interest?

36. What is the present worth of $3525, due in 3 years; discount being allowed as in the last example?

37. How much must be advanced to discharge a debt of $700, due in 8 years; discounting at the rate of 5 per cent. compound interest?

38. What is the present worth of $1000, due in 20 years; discounting at the rate of 6 per cent. compound interest? How much is the discount?

XXXIV.

ANNUITIES.

An ANNUITY is a fixed sum of money payable periodically, for a certain length of time, or during the life of some person, or for ever.

Although the term *annuity*, in its proper sense, applies only to annual payments, yet payments which are made semiannually, quarterly, monthly, &c., are also called annuities.

Pensions, salaries, and rents, come under the head of annuities. Annuities may, however, be purchased by the present payment of a sum of money. The party selling annuities, is usually an incorporated trust company, instituted and regulated upon principles similar to those of an insurance company. The company has an office, called an *annuity office*, where all its business is transacted.

The present worth of an annuity which is to continue for ever, is that sum of money, which would yield an interest equal to the annuity. But the present worth of an annuity which is to terminate, is a sum, which, being put on compound interest, would, at the termination of the annuity, amount to just as much as the payments of the annuity would amount to, provided they should severally be put on compound interest, as they became due.

The sum to be paid for the purchase of a life annuity—which is the same as its present worth—depends not only upon the rate of interest, but, also upon the probable continuance of the life or lives on which the annuity is granted. In order to bring data of this kind into numbers, the bills of mortality in different places have been examined, and from them, tables have been constructed, which show how many persons, upon an average, out of a certain number born, are left alive at the end of each year; and from these tables others have been constructed, showing the expected continuance of human life, at every age, according to probabilities. We shall not, however, treat the subject of *life* annuities in this work, and would refer readers, who wish to become thoroughly acquainted with its theory, to the writings of Simpson, De Moivre, Bailey, Price, and Milne.

PROBLEM I. To find the amount of an annuity, which has been forborn for a given time.

Before presenting the rule, let us inquire what would be the amount of an annuity of $100, forborn 4 years, allowing 5 per cent. compound interest? The last year's payment will, obviously, be $100 without interest; the last but one will be the amount of $100 for 1 year; the last but two will be the amount of $100 for 2 years; and so on: and the sum of the amounts will be the answer. Now the last payment with the amounts for the several years, form a series of *continual proportionals.* We, therefore, adopt the following—

RULE. *Find the sum of an increasing series of continual proportionals, whose first term is the annuity, whose ratio is the amount of 1 dollar for 1 year, and whose number of terms is the number of years. This sum is the amount.* See ART. XXXIII, Problems 1st and 2nd.

1. What is the amount of an annuity of $200, which has been forborn 14 years; allowing 6 per cent. interest?

2. What is the amount of an annuity of $50, which has been forborn 20 years; interest being 5 per cent.?

3. What is the amount of an annual rent of $150, forborn 7 years; allowing interest at 5 per cent.?

4. If the annual rent of a house be in arrears 4 years, what is the amount, allowing 10 per cent. interest?

5. Suppose a person, who has a salary of $600 a year, payable quarterly, to allow it to remain unpaid for 3 years, how much would be due him; allowing quarterly compound interest at 6 per cent. per annum?

6. What is due on a pension of $150 a year, payable half-yearly, but forborn 2 years; allowing half-yearly compound interest, at 4½ per cent. per annum?

7. What is due on a pension of $300 a year, payable quarterly, but forborn 2½ years; allowing quarterly compound interest, at 5 per cent. per annum.

PROBLEM II. To find the present worth of an annuity which is to terminate in a given number of years.

Before giving the rule, let us inquire, what is the present worth of an annuity of $100, to continue 4 years, allowing 5 per cent. interest? The present worth is, obviously, a sum, which, at compound interest, would produce an amount equal to the *amount of the annuity*. Now we can find the amount of any sum at compound interest, by multiplying the sum by the amount of 1 dollar for a year, as many times as there are years. Hence, to find a sum which will produce a given amount in a given time, we must reverse the process, and *divide* by the amount of 1 dollar for the time. Applying this principle to the example in question, we find by the preceding rule, that the amount of the annuity is $431. Then, dividing this amount by the amount of 1 dollar for 4 years, we find the present worth to be $354.593+

RULE. *Find the amount of the annuity as if it were in arrears for the whole time, and divide this amount by the amount of 1 dollar at compound interest for the same time; the quotient will be the present worth.*

8. What is the present worth of an annuity of $500, to continue 10 years; interest being 6 per cent.?

9. What is the present worth of an annuity of $80, to continue 22 years; interest being 5 per cent.?

The operations in this rule being tedious, we introduce, upon the next page, a table, showing the present worth of $1 annuity, at 4, 5, 6, and 7 per cent., for every number of years, from 1 to 30. To find the present worth of an annuity by the use of this table, *multiply the present worth of 1 dollar for the number of years, by the annuity.*

Y'rs.	4 per cent.	5 per cent.	6 per cent.	7 per cent.
1	.9615	.9523	.9433	.9345
2	1.0860	1.8594	1.8333	1.8080
3	2.7750	2.7232	2.6730	2.6243
4	3.6298	3.5459	3.4651	3.3872
5	4.4518	4.3294	4.2123	4.1001
6	5.2421	5.0756	4.9173	4.7665
7	6.0020	5.7863	5.5823	5.3892
8	6.7327	6.4632	6.2097	5.9712
9	7.4353	7.1078	6.8016	6.5152
10	8.1109	7.7217	7.3600	7.0235
11	8.7605	8.3064	7.8868	7.4986
12	9.3850	8.3632	8.3838	7.9426
13	9.9856	9.3935	8.8526	8.3576
14	10.5631	9.8986	9.2949	8.7454
15	11.1184	10.3796	9.7122	9.1079
16	11.6523	10.8377	10.1058	9.4466
17	12.1656	11.2740	10.4772	9.7632
18	12.6593	11.6895	10.8276	10.059
19	13.1339	12.5883	11.1581	10.335
20	13.5903	12.4622	11.4699	10.594
21	14.0291	12.8211	11.7640	10.835
22	14.4511	13.1630	12.0415	11.061
23	14.8568	13.4885	12.3033	11.272
24	15.2469	13.7986	12.5503	11.469
25	15.6220	14.0939	12.7833	11.653
26	15.9827	14.3751	13.0031	11.825
27	16.3295	14.6430	13.2105	11.986
28	16.6630	14.8981	13.4061	12.137
29	16.9837	15.1410	13.5907	12.277
30	17.2920	15.3724	13.7648	12.409

10. What is the present worth of an annuity of $21.54, for 7 years; interest being 6 per cent.?

11. What is the present worth of an annuity of $936, for 20 years, at 5 per cent.?

12. What is the present worth of an annuity of $258, for 17 years, at 4 per cent.?

13. Find the present worth of an annuity of $796.50, to continue 28 years; interest being 7 per cent.?

14. A young man purchases a farm for $924; and agrees to pay for it in the course of 7 years, paying $\frac{1}{7}$ part of the price at the end of each year. Allowing interest to be 6 per cent., how much cash in advance will pay the debt?

15. Allowing interest to be 5 per cent., which will be in my favor, to pay $15 a year for 10 years, or, to pay $160 in advance?—by how much?

When an annuity does not commence until a given time has elapsed, or some particular event has taken place, it is called a REVERSION.

PROBLEM III.   To find the present worth of an annuity in reversion.

RULE.   *Find, (by Problem 2nd.), the present value of the annuity from the present time till the end of the period of its continuance: find, also, its value for the time before it is to commence: the difference of these two results will be the present worth.*

16. What is the present worth of an annuity of $200, to be continued 7 years, but not to commence till 2 years hence; interest being 6 per cent.?

17. Find the present worth of a reversion of $152 a year, to commence in 6 years, and to continue 18 years; interest being 4 per cent.

18. What is the present worth of a reversion of $75 a year, to commence in 5 years, and to continue 24 years; interest being 6 per cent.?

19. What must be paid for the purchase of a reversion of $450 a year, to commence in 5 years, and to continue 13 years; interest being 5 per cent.?

20. Find the present worth of a reversion of $942.30 a year, to commence in 2 years, and to continue 11 years; interest being 7 per cent.

21. A father leaves to his son, a rent of $310 per annum, for 8 years, and, the reversion of the same rent to his daughter for 14 years thereafter. What is the present worth of the legacy of each?

22. What is the present worth of a reversion of $100 a year, to commence in 4 years, and to continue for ever; interest being 6 per cent.?

This annuity continuing for ever, will, when it commences, be worth that sum of money which would yield $100 a year, at 6 per cent. interest. Therefore, after finding the principal, whose interest is $100 per annum, deduct from it a compound discount for 4 years; the remainder will be the present worth.

23. What is the present worth of a reversion of $824 a year to commence in 7 years, and to continue for ever; interest being 5 per cent.?

24. What is the present worth of a reversion of $530 a year, to commence in 22 years, and to continue for ever; interest 7 per cent.?

25. How much must be paid, at present, for a share in a fund, which, after the lapse of 20 years, will yield an income of $400 a year; interest 6 per cent.?

26. How much must be paid, at present, for the title to an annuity of $1000, to commence in 40 years; interest being 5 per cent.?

XXXV.

ALLIGATION.

ALLIGATION relates to finding the mean value of a mixture composed of several ingredients of different values, and is considered under two heads, viz. Alligation Medial, and Alligation Alternate.

ALLIGATION MEDIAL.

We rank under the head of Alligation medial, those questions, in which the several ingredients and their respective values are given, and the mean value of the compound is required.

For example, a wine merchant bought several kinds of wine, as follows; 160 gallons at 40 cents per gallon; 75

gallons at 60 cents per gallon; 225 gallons at 48 cents per gallon; 40 gallons at 85 cents per gallon; and mixed them together. It is required to find the cost of a gallon of the mixture.

Now, if we find the whole cost of the several kinds of wine, and divide it by the whole number of gallons, it is evident, that the quotient will be the cost of a single gallon of the mixture.

160 gallons, at 40 cents per gal., cost	$ 64.00
75 gallons, at 60 cents per gal., cost	$ 45.00
225 gallons, at 48 cents per gal., cost	$108.00
40 gallons, at 85 cents per gal., cost	$ 35.00

500 the whole number of gallons, cost $252.00

$252.00 ÷ 500 = .504, or 5 cents and 4 mills.

Therefore, to find the mean value of a compound, composed of several ingredients, of different values, we give the following

RULE. *Find the value of each ingredient, add these values together, and divide their sum by the sum of the ingredients. The quotient is the mean value.*

1. A farmer mixed together 5 bushels of rye worth 70 cents a bushel, and 10 bushels of corn worth 60 cents a bushel, and 5 bushels of wheat worth $1.10 a bushel. What is a bushel of the mixture worth?

2. A grocer mixed together 38 lb. of tea at 50 cents a pound, $15\frac{1}{4}$ lb. at 80 cents, $12\frac{1}{2}$ lb. at 60 cents, $8\frac{3}{4}$ lb. at 96 cents, $77\frac{1}{2}$ lb. at 32 cents, and sold the mixture at a profit of 20 per cent. At what price per pound did he sell it?

3. A goldsmith melts together 11 ounces of gold 23 carats fine, 8 ounces $21\frac{1}{4}$ carats fine, 6 ounces of pure gold, and 2 ounces of alloy. How many carats fine is the mixture?

We remark, that a carat is a 24th part. Thus, 23 carats fine, means $\frac{23}{24}$ of pure metal. Pure gold is $\frac{24}{24}$. Alloy is considered of no value.

4. On a certain day, the mercury in the thermometer was observed to stand, 2 hours at 60 degrees, 3 hours at

52°, 4 hours at 64°, 3 hours at 67°, 1 hour at 72°, and 1 hour at 75°. What was the mean temperature for that day?

5. A dealer bought 24½ gallons of sirop at 34 cents a gallon, and 24½ gallons at 38 cents a gallon, and mixed both quantities and 14 gallons of water together, and sold the mixture at a profit of 50 per cent. At what price per gallon did he sell it?

6. A goldsmith melts together 3 ounces of gold 18 carats fine, 2 ounces 21 carats fine, and 1 ounce of pure gold. What is the fineness of the compound?

ALLIGATION ALTERNATE.

Under the head of Alligation Alternate are included those questions, in which the respective rates of the different ingredients are given, to compose a mixture of a fixed rate. It is the reverse of Alligation Medial, and may be proved by it.

If we would find what quantities of two ingredients, different in value, would be required to make a compound of a fixed value, it is evident, that, when the value of the required compound exceeds that of one ingredient just as much as it falls short of the value of the other, we must take equal quantities of the ingredients to make the compound; because there is just as much lost on the one, as is gained on the other.

If the value of the compound exceeds that of one ingredient twice as much as it falls short of the value of the other, we must take of the ingredients in the ratio of ½ to 1, or 1 to 2. For instance, if we would mix wines, at 4 dollars and 1 dollar a gallon, in such proportion that the mixture should be worth 2 dollars a gallon, we must take 1 gallon at 4 dollars to 2 gallons at 1 dollar; because there is just as much lost on 1 gallon at 4 dollars, as is gained on 2 gallons at 1 dollar.

If we would mix wines, at 6 dollars and 2 dollars a gallon, in such proportion as would make the mixture worth 3 dollars a gallon, we should take of the two kinds in the ratio of $\frac{1}{2}$ to $\frac{1}{6}$, or 1 to 3; for, in this instance, there is as much lost on 1 gallon at 6 dollars, as is gained on 3 gallons at 2 dollars.

We see by the preceding ratios, that the nearer the value of the mixture is to that of one of the ingredients, the greater must be the relative quantity of this ingredient, in forming the compound; and the farther the value of the mixture is from that of one of the ingredients, the less must be the relative quantity of this ingredient in making the compound.

Hence, if we make the difference between the rate of each ingredient and that of the compound, the denominator of a fraction having 1 for its numerator, these fractions express the ratio of the ingredients required to make the compound; and, when these fractions are reduced to a common denominator, the numerators express the required ratio of the ingredients.

If, for example, it be required to mix gold of 12 carats fine with gold of 22 carats fine, in such proportion that the mixture may be 18 carats fine, we can ascertain the proportion of each kind in the following manner. The difference between 18 and 12 is 6; making 6 the denominator of a fraction with 1 for its numerator, we have the fraction $\frac{1}{6}$; taking the difference between 18 and 22, we in like manner obtain the fraction $\frac{1}{4}$; therefore, the fractions, $\frac{1}{6}$ and $\frac{1}{4}$, express the required proportion of each sort of gold. These fractions, when reduced to a common denominator, are $\frac{4}{24}$ and $\frac{6}{24}$, and the numerators express the required proportion of each sort. Therefore, we must take 4 grains of 12 carats fine, and 6 grains of 22 carats fine; or, in that ratio.

If, for a second example, we would make a mixture 18 carats fine from gold of 15 carats and 20 carats fine, we should, in the same manner, obtain the fractions, $\frac{2}{6}$ and $\frac{3}{6}$, to express the required proportion of the two sorts of gold; consequently, in this instance, we should take 2 grains of 15 carats fine, and 3 grains of 20 carats fine.

Therefore, since the fineness of the compound is the same in both the preceding examples, if we would make a compound 18 carats fine, from the four kinds of gold mentioned in the two examples, we should take 4 grains of 12 carats, 6 grains of 22 carats, 2 grains of 15 carats, and 3 grains of 20 carats fine.

Now, these results may be readily obtained by writing the rates of the given simples one under another, in regular order, beginning either with the least or greatest, and alligating one of a less with one of a greater rate than that of the compound, and writing the difference between the rate of each simple and the rate of the compound, against the rate of the simple with which it is alligated.

Thus, 18 | 12 — 4 grains 12 carats fine
         | 15 — 2 "   15 "   "
         | 20 — 3 "   20 "   "
         | 22 — 6 "   22 "   "

We may connect the rates of the simples differently, and obtain equally correct, but different results.

Thus, 18 | 12 — 2
          | 15 — 4
          | 20 — 6
          | 22 — 3

It must be observed, that the two simples linked together, must always be one of a less, and the other of a greater rate, than the rate of the compound.

By connecting a less rate with a greater, and placing the differences between them and the mixture rate alternately, the gain on the one is precisely balanced by the loss on the other. This being true of every two, it is true of all the simples in the question, whatever may be their number.

It is obvious, that a question in Alligation Medial admits of a great variety of answers, all agreeing with the requisition of the question; for we may variously alligate the values of the ingredients, and thus obtain various results, all of which will be correct; and we may add all these together, and the results will be correct answers. We may also multiply, or divide the quantities found; for, if two quantities of two simples make a balance of loss and gain in relation to the value of the compound, so must also the double or treble, the half or third part, or any other ratio of the quantities.

18

We shall give the questions in Alligation Alternate under four cases.

CASE I.   The ratios of the several ingredients being given, to make a compound of a fixed rate.

RULE.   First—*Write the rates of the several ingredients in a column under one another.*

2dly—*Connect with a continued line the rate of each ingredient less than the rate of the compound, with one or more rates greater than the rate of the compound; and each of a greater rate than the rate of the compound with one or more of a less rate.*

3dly—*Write the difference between the rate of each ingredient and the rate of the compound, opposite the rate of the ingredient with which it is connected.*

4thly—*If only one difference stand against any rate, it will be the required quantity of the ingredient of that rate; but, if there be several, their sum will be the quantity required.*

7. A goldsmith has gold of 17, 18, and 22 carats fine, and also pure gold.   What proportion of each sort must he take, to compose a mixture 21 carats fine ?

8. Having gold of 12, 16, 17, and 22 carats fine, what proportion of each kind must I take, to make a compound 18 carats fine ?

9. A merchant has spices at 30, 33, 67, and 86 cents a pound.   How much of each sort must he take, to make a mixture worth 56 cents a pound ?

10. A wine merchant has Canary wine at 50 cents a gallon, Sherry at 76 cents, and Claret at 175 cents per gallon.   How much of each sort must he take, to make a mixture worth 87 cents a gallon ?

11. A goldsmith wishes to mix gold of 16, 18, 19, and 23 carats fine, with pure gold, in such proportions that the composition may be 20 carats fine.   What quantity of each must he take ?   .

12. It is required to mix different sorts of wine, at 56, 62, and 75 cents per gallon, with water, in such proportions that the mixture may be worth 60 cents a gallon. How much of each must be taken ?

13. How much corn at 52 cents a bushel, rye at 56 cents, wheat at 90 cents, and wheat at 1 dollar a bushel, must be mixed together, that the composition may be worth 62 cents a bushel ?

14. A silversmith wishes to mix alloy with silver of 10, and 7 ounces fine, and pure silver, in such proportion that the mass may be 9 ounces fine: 12 ozs. fine being pure. How much of each must he take?

CASE II. When one of the ingredients is limited to a certain quantity.

RULE. *Find the quantity of each ingredient, as in Case 1st. in the same manner, as though no such limitation were made; then as the difference against that simple, whose quantity is given, is to each of the other differences, so is the given quantity of that simple to the quantity required of each of the other simples.*

15. A trader has 90 pounds of tea worth 40 cents a pound, which he would mix with some at 50 cents, some at 85 cents, and some at 90 cents. How much of each of the other sorts must he mix with the 90 pounds, to make a mixture worth 60 cents a pound ?

First solution.

$$60 \begin{vmatrix} 40 & 30 \\ 50 & 25 \\ 85 & 10 \\ 90 & 20 \end{vmatrix}$$

thus 30 : 25 = 90 : 75
30 : 10 = 90 : 30
30 : 20 = 90 : 60

Ans. 75 lb. at 50 cents, }
30 lb. at 85 cents, and }
60 pounds at 90 cents. }

Second solution.

$$60 \begin{vmatrix} 40 & 25 \\ 50 & 30 \\ 85 & 20 \\ 90 & 10 \end{vmatrix}$$

thus 25 : 30 = 90 : 108
25 : 20 = 90 : 72
25 : 10 = 90 : 36

Ans. 108 lb. at 50 cents,
72 lb. at 85 cents, and
36 pounds at 90 cents.

16. A farmer wishes to mix corn at 54 cents a bushel, rye at 61 cents a bushel, and wheat at 96 cents a bushel, with 3 bushels of wheat worth 1 dollar and 10 cents a bushel. How much of each of the other three must be mixed with the 3 bushels of wheat at 1 dollar and 10 cents a bushel, that the mixture may be worth 75 cents a bushel ?

17. How much gold of 16, 20, and 24 carats fine, and how much alloy, must be mixed with 10 ounces of 18 carats fine, that the composition may be 22 carats fine?

18. How much silver of 6.5 ounces fine, and of 10.5 ounces fine, and alloy, must be mixed with 17.1 ounces of pure silver, that the mass may be 9.5 oz. fine?

It must be observed, that pure silver is 12 ounces fine.

CASE III. When two or more of the ingredients are limited in quantity.

RULE. *Find, as in Alligation Medial, what will be the rate of a mixture made of the given quantities of the limited ingredients only; then consider this as the rate of a limited ingredient, whose quantity is the sum of the quantities of the limited ingredients, from which, and the rates of the unlimited ingredients, proceed to calculate the several quantities required, as in* Case II.

19. I have 18 gallons of wine at 48 cents a gallon, 8 gallons at 52 cents, and 4 gallons at 85 cents, and would mix the whole with two other kinds of wine, one at a dollar and 26 cents, the other at 2 dollars and 12 cents a gallon. How much of the wine at a dollar and 26 cents, and of that at 2 dollars and 12 cents, must I mix with the other three, that the mixture may be worth a dollar a gallon?

     18 gal. at .48 come to $ 8.64
      8 gal. at .52   "   4.16
      4 gal. at .85   "   3.40

The 30 gal. come to $16.20, which is .54 a gallon. 54 cents a gallon being the mean value of the 30 gallons, contained in the three kinds that are limited, I must now inquire how much of each of the other two sorts of wine at 1 dollar 26 cents, and 2 dollars 12 cents, must be mixed with 30 gallons at 54 cents a gallon, to make a mixture worth one dollar a gallon.

$$100 \begin{vmatrix} 54 \\ 126 \\ 212 \end{vmatrix} \begin{matrix} 26 + 112 = 138 \\ - \;-\;-\;- 46 \\ - \;-\;-\;- 46 \end{matrix}$$

gal.   gal.   gal.   gal.
Now as 138 : 46 = 30 : 10

Therefore, I must take 10 gallons each of the two sorts, which are worth 1 dollar 26 cents, and 2 dollars 12 cents a gallon.

20. How much gold, of 14 and 16 carats fine must be mixed with 6 ounces of 19, and 12 oz. of 22 carats fine, that the composition may be 20 carats fine?

21. A silversmith has silver of 6, 7, and 9 ounces fine, which he wishes to mix with 9 ounces of 10 ounces fine, and 9 ounces of pure silver, to make a mass, that shall be 8 ounces fine. How much of each of the three first must he take?

22. A lady purchases 7 yards of calico at 22 cents a yard, and 7 yards at 20 cents a yard, and wishes to know how many yards of two other kinds, one at 16 cents and the other at 17 cents a yard, she must purchase, to make the average price of the whole 18 cents a yard. Find the two quantities.

CASE IV. When the whole compound is limited to a certain quantity.

RULE. *Find an answer, as in* Case I, *by alligating; then, as the sum of the quantities thus found, is to the given quantity, so is the quantity of each ingredient found by alligating, to the required quantity of it.*

23. A goldsmith has gold of 15, 17, 20, and 22 carats fine; and would melt together of all these sorts so much, as to make a mass of 40 ounces 18 carats fine. How much of each sort is required?

$$
18\begin{vmatrix}15 & 4 \\ 17 & 2 \\ 20 & 1 \\ 22 & 3\end{vmatrix}
\qquad \text{Or thus} \qquad
18\begin{vmatrix}15 & 2 \\ 17 & 4 \\ 20 & 3 \\ 22 & 1\end{vmatrix}
$$

$\overline{10}:40=4:16$	$\overline{10}:40=2:\ 8$
10 : 40 = 2 : 8	10 : 40 = 4 : 16
10 : 40 = 1 : 4	10 : 40 = 3 : 12
10 : 40 = 3 : 12	10 : 40 = 1 : 4

Ans. 16 oz. of 15; 8 oz. of 17; 4 oz. of 20; and 12 oz. of 22 carats fine.

24. Having three sorts of raisins at 9, 12, and 18 cents

18*

a pound, what quantity of each sort must I take, to fill
a cask containing 210 pounds, that its contents may be
worth 14 cents a pound?

25. Of four different kinds of apples at 31, 37, 46, and
74 cents a bushel, what quantity of each must be taken,
to fill a bin containing 9 bushels, to make its contents
worth 50 cents a bushel?

XXXVI.

PERMUTATIONS.

PERMUTATION—which is also called *variation*—means
the different ways in which the order or relative position
of any given number of things may be changed. The
only object to be regarded in Permutation, is *the order in
which the things are placed;* for no two arrangements are
to have all the quantities in the same relative position.

For example, two things, a, and b, are capable of only
two changes in their relative position, viz. ab, ba; and
this number of changes is expressed by 1×2; but three
things, a, b, and c, are capable of six variations, viz.
a b c, a c b, b a c, b c a, c a b, c b a, and this number
of permutations is expressed by $1 \times 2 \times 3$; and four things,
a, b, c, and d, are capable of 24 variations, viz. a b c d,
a b d c, a c b d, a c d b, a d b c, a d c b; b a c d, b a
d c, b c a d, b c d a, b d a c, b d c a; c a b d, c a d b,
c b a d, c b d a, c d a b, c d b a; d a b c, d a c b, d b
a c, d b c a, d c a b, d c b a; and this number of permutations is expressed by $1 \times 2 \times 3 \times 4$.

In like manner, when there are 5 things, every four of
them, leaving out the 5th, will have 24 variations; consequently by taking in the 5th, there will be 5 times 24
variations.

PROBLEM I. To find the number of permutations that
can be made of any given number of things, all different
from each other.

RULE. *Multiply the terms of the natural series of*

numbers, from 1 up to the given number of things, continually together, and the product will be the answer.

1. How many changes can be made in the order of the six letters, a b c d e f ?

2. How many changes may be rung on seven bells?

3. Five gentlemen agreed to board together, as long as they could seat themselves every day in a different position at the dinner table. How long did they board together?

4. How many changes may be made in the order of the words in the following verse? Proci tot tibi sunt, virgo, quot sidera coelo.

5. How many different sums of dollars can be expressed by the nine digits, without using any one of them more than once in the same sum?

6. How many different arrangements may be made in seating a class of 20 scholars?

7. A gentleman, who had a wife and eight daughters, one day said to his wife, that he intended to arrange the family in a different order every day at the dinner table, and that he would never give one of his daughters in marriage, till he had completed all the different arrangements of which the family was capable. How many years from that day must elapse, before either of his daughters can be married?

When several of the things are of one sort, and several of another, &c. the changes that can be made upon the whole is not so great, as when all the things are different. For instance, we have seen that the letters a b c admit of six variations; but, if two of the quantities be alike, as a a b, the six variations are reduced to three, a a b, b a a, a b a, which may be expressed by $\frac{1 \times 2 \times 3}{1 \times 2}$. We have also seen that the letters a b c d admit of 24 variations; but if we have a a b b, the 24 variations are reduced to six, viz. a a b b, a b b a, a b a b, b b a a, b a a b, b a b a, and this number of variations may be expressed by $\frac{1 \times 2 \times 3 \times 4}{1 \times 2 \times 1 \times 2}$. Hence, we have, as follows,—

PROBLEM II. To find the number of changes that may be made in the arrangement of a given number of things, whereof there are several things of one sort, several of another, &c.

RULE. *Take the natural series of numbers from 1 up to the given number of things, as if they were all different, and find the product of the terms.*

Then take the natural series from 1 up to the number of similar things of one sort, and the same series up to the number of similar things of a second sort, &c., and divide the first product by the joint product of all these series, and the quotient will be the answer.

8. Find how many changes can be made in the order of the letters a a a b b c.

If the letters in this question were all different, they would admit of $1 \times 2 \times 3 \times 4 \times 5 \times 6 = 720$ variations; but since a is found 3 times, we must divide that number of variations by $1 \times 2 \times 3$; and, since b occurs twice, we must again divide by 1×2; therefore the number of variations will be $\frac{1 \times 2 \times 3 \times 4 \times 5 \times 6}{1 \times 2 \times 3 \times 1 \times 2} = 60$.

9. How many changes can be made in the order of the letters a a a b b b b c c d e e?

10. How many variations may take place in the succession of the following musical notes, fa, fa, fa, sol, sol, la, mi, fa?

11. How many whole numbers can you make out of the number 1220055055, using all the figures each time?

12. How many variations can be made in the order of the figures in the number 97298279289?

PROBLEM III. Any number of different things being given, to find how many changes can be made out of them, by taking a given number of the things at a time.

RULE. *Take a series of numbers commencing with the given number of things and decreasing by 1, till the number of terms is equal to the number of things to be taken at a time, and the product of all the terms of this series will be the answer.*

To illustrate the rule, we will take the four letters a b c d, and find the number of variations that can be made upon them, by taking two at a time. In the first place, we will write the letter *a* on the left hand of each of the other letters, and the variations will be three, viz. a b, a c, a d; we will do the same with each of the other letters, thus, b a, b c, b d; c a, c b, c d; d a, d b, d c. Now we have all the changes that can be made upon the four letters, taking two at a time, and they are $4 \times 3 = 12$.

We will also find, in the same manner, how many changes can be made on the same four letters, by taking three at a time; writing *a* on the left, thus, a b c, a b d; a c b, a c d; a d b, a d c, we have $3 \times 2 = 6$ variations. Now, since each of the letters is to be written in the same manner on the left, we shall have four such classes of variations, and the whole number will be $4 \times 3 \times 2 = 24$ variations.

13. How many changes can be made upon the letters a b c d e f, by taking three at a time?

14. How many different whole numbers can be expressed by the nine digits, by using two at a time?

15. How many different whole numbers can be expressed by the nine digits, by using four at a time?

16. How many different numbers can you express with the nine digits and a cipher, by using five at a time?

XXXVII.

COMBINATIONS.

COMBINATION consists in taking a less number of things out of a greater without any regard to the order in which they stand. This is sometimes called Election or Choice.

No two combinations can have the same quantities; for instance, the quantities, *a* and *b*, admit of only one combination, because a b and b a are composed of the same quantities; but, if a third quantity *c* be added, we can make three combinations of two quantities out of them, because the third quantity *c* may be added to each of the two former, thus, a b, a c, b c; this number of combinations may be expressed by $\frac{3 \times 2}{1 \times 2}$. If we add a fourth letter,

d, we can make six combinations of two letters out of the four, since the new quantity d, may be combined with each of the former ones; thus, a b, a c, b c, a d, b d, c d; and this number of combinations may be expressed by $\frac{4\times3}{1\times2}$.

If we would make a combination of four, it is evident that only one such combination can be made out of the letters a b c d; but if a fifth letter, e, be added, we can make five such combinations; thus, a b c d, a b c e, a b e d, a e c d, b c d e; and this number of combinations may be expressed by $\frac{5\times4\times3\times2}{1\times2\times3\times4}$.

PROBLEM I.   To find the number of combinations from any given number of things, all different from each other, taking a given number at a time.

RULE.   *Take a series of numbers, the first term of which is equal to the number of things out of which the combinations are to be made, and decreasing by 1, till the number of terms is equal to the number of things to be taken at a time, and find the product of all the terms.*

Then take the natural series 1, 2, 3, &c. up to the number of things to be taken at a time, and find the product of all the terms of this series.

Divide the former product by the latter, and the quotient will be the answer.

1. How many combinations of 3 letters can be made out of the 6 letters a b c d e f?

2. How many different yoke of oxen may be selected from twelve oxen?

3. How many different span of horses can be selected from eighteen horses?

4. A drover agreed with a farmer for a dozen sheep, to be selected out of a flock of two dozen; but while he was making the selection, the farmer told him, he might take the whole flock, if he would give him a cent for every different dozen that could be selected from it. To this the drover readily agreed. How many dollars did the whole flock come to, at that rate?

5. A general, who had often been successful in war,

was asked by his king, what reward he should confer upon him for his services. The general only desired a farthing for every file, of 10 men in a file, which he could make with a body of 100 men. How much did the general's modest request amount to ?

PROBLEM II. To find the various combinations of a given number of things, which may be made out of an equal number of sets of different things, one from each set.

RULE. *Multiply the number of things in the several sets continually together, and the product will be the answer.*

A combination of this kind is called the composition of quantities. The rule may be illustrated thus. If there are only two sets, and we combine every quantity of one set with every quantity of the other set, we shall make all the compositions of two things in these two sets; and the number of compositions is evidently the product of the number of things in one set by the number of things in the other set. Again, if there are three sets, then the compositions of two in any two of the sets, being combined with every quantity of the third set, will make all the compositions of three in the three sets. That is, the compositions of two in any two of the sets, being multiplied by the number of things in the third set, will give all the compositions of three in the three sets; and this result is the joint product of all the numbers in the three sets.

6. Suppose there are four companies, in each of which there are 9 men; in how many ways can 4 men be chosen, one out of each company ?

7. Suppose there are five parties, at one of which there are 6 young ladies, at another 8, at a third 5, at a fourth 7, at a fifth 10. How many choices are there, in selecting 5 young ladies, one from each party?

8. How many changes are there in throwing four dice, each die having six sides ?

9. A certain farmer has 5 barns, in one of which he has 15 cows, in another 11, in another 5, in another 9,

and in another 7. · How many different selections may be made, in choosing 5 cows, one from each barn?

10. How many variations can be made in selecting a flock of a dozen sheep from 12 folds, one from every fold, in each of which there are 10 sheep?

11. In a certain school there are seven classes, the first containing 12 boys, the second 7, the third 9, the fourth 10, the fifth 11, the sixth 8, and the seventh 13. How many variations can be made in selecting 7 boys, one from each class?

XXXVIII.

EXCHANGE.

Scholars, who are to prosecute a course of classical studies, and those, who are not expected to engage in any extensive mercantile business, may omit the exercises in this article.

EXCHANGE is the act of paying or receiving the money of one country for its equivalent in the money of another country, by means of *Bills of Exchange*. This operation, therefore, comprehends both the reduction of moneys and the negotiation of bills; it determines the comparative value of the currencies of different nations, and shows how foreign debts are discharged, and remittances made from one country to another, without the risk, trouble, or expense of transporting specie or bullion.

A Bill of Exchange is a written order for the payment of a certain sum of money, at an appointed time. It is a mercantile contract, in which four persons are mostly concerned, as follows.

First—The *Drawer*, who receives the value, and is also called the *maker* and *seller* of the Bill.

Second—The debtor in a distant place, upon whom the Bill is drawn, and who is called the *Drawee*. He also is called the *Acceptor*, after he accepts the Bill, which is an engagement to pay it when due.

Third—The person who gives the value for the Bill, who is called the *Buyer*, *Taker* and *Remitter*.

. Fourth—The person to whom the bill is ordered to be paid, who is called the *Payee*, and who may, by endorsement, pass it to any other person.

Many mercantile payments are made in Bills of Exchange, which pass from hand to hand, until due, like any other circulating medium; and the person who at any time has a Bill in his possession, is called the *holder*.

To transfer a Bill payable to order, the *payee* should express his order of paying to another person, which is always done by an endorsement on the back of the Bill.

An endorsement may be blank or special. A blank endorsement consists only of the endorser's name, and the Bill then becomes transferable by simple delivery. A special endorsement orders the money to be paid to a particular person, who is called the *endorsee*, who must also endorse the Bill, if he negotiates it. A blank endorsement may always be filled up with any person's name, so as to make it special. Any person may endorse a Bill, and every endorser, as well as the acceptor, is a security for the Bill, and may be sued for payment.

In reckoning when a Bill, payable after *date*, becomes due, the day on which it is dated, is not included. When the time is expressed in months, calendar months are understood; and when a month is longer than the succeeding, it is a rule not to go, in the computation, into a third month. Thus, if a Bill be dated the 28th, 29th, 30th, or 31st, of January, and payable one month after date, the term equally expires on the last day of February.

An endorsement may take place at any time after the Bill is issued, even after the day of payment is elapsed.

When the holder of a Bill dies, his executors may endorse it; but, by so doing, they become answerable to their endorsee personally, and not as executors.

A Bill payable to bearer is transferred by simple delivery, without any endorsement.

Bills should be presented for acceptance, as well as for payment, during the usual hours of business.

The common way of accepting a Bill is for the drawee to write his name at the bottom or across the body of it, with the word, *accepted*.

When acceptance or payment has been refused, the Holder of the Bill should give regular and immediate notice to all the parties, to whom he intends to resort for payment; for if he do not, they will not be liable to pay.

With respect to the manner, in which notices of non-acceptance or non-payment are to be given, a difference exists between Inland and Foreign Bills.

In the case of Foreign Bills, a Protest is indispensably necessary: thus, a Public Notary appears with the Bill, and demands either acceptance or payment (as the case may be;) and on being refused, he draws up an instrument, called a Protest, expressing that acceptance or payment (as the case may be) has been demanded and refused, and that the holder of the Bill intends to recover any damages which he may sustain in consequence. This instrument is admitted, in foreign countries, as a legal proof of the fact.

The Protest on a Foreign Bill should be sent as soon as possible, to the *drawer* or *negotiator;* and if it be for non-payment, the Bill must be sent with the Protest.

A Protest is not absolutely necessary to entitle the *holder* to recover the amount of an Inland Bill from the *drawer* or *endorser:* it is sufficient if he give notice, by letter or otherwise, that acceptance or payment (as the case may be) has been refused, and that he does not mean to give credit to the drawee.

If the person, who is to accept, has absconded, or cannot be found at the place mentioned in the Bill, Protest is to be made, and notice given, in the same manner as if acceptance had been refused.

It is customary, as a precaution against accident or miscarriage, to draw three copies of a Foreign Bill, and to send them by different conveyances. They are denominated the *First, Second,* and *Third of Exchange;* and when any one of them is paid, the rest become void.

When acceptance is refused, and the Bill is returned by Protest, an action may be commenced immediately against the *Drawer,* though the regular time of payment be not arrived. His debt, in such case, is considered as contracted the moment the Bill is drawn.

FORM OF A BILL OF EXCHANGE.

New York, Nov. 4, 1834.

EXCHANGE *for £3000 sterling.*

At thirty days sight of this, my first of Exchange, (second and third of the same tenor and date not paid), pay to Robert N. Foster, or order, Three Thousand Pounds Sterling, with or without further advice from me.

EDWIN D. HARPER.

Messrs. KNOX *and* FARNHAM,
Merchants, London.

INLAND EXCHANGE relates only to remitting Bills from one commercial place to another in the same country; by which means debts are discharged more conveniently than by cash remittances.

Suppose, for example, A of New Orleans is creditor to B of Boston 1000 dollars, and C of New Orleans is debtor to D of Boston 1000 dollars; both these debts may be discharged by means of one Bill. Thus, A draws for this sum on B, and sells his Bill to C, who remits it to D, and the latter receives the amount, when due, from B. Thus, by a transfer of claims, the New Orleans debtor pays the New Orleans creditor, and the Boston debtor the Boston creditor, and no money is sent from one place to the other. This business is usually conducted through the medium of Banks, which are in the habit of buying both foreign and inland Bills of exchange, and transmitting them to the places on which they are drawn for acceptance.

Inland Bills of Exchange are sometimes called *Drafts*; and the following short form of the instrument is adopted.

560 $\frac{50}{100}$ *Boston, Dec. 8, 1834.*

Three months after date, pay to the order of Charles S. Hooper, Five hundred dollars 50 cents, value received, and charge the same to account of

HANNUM & LORING.

To STEPHEN FROTHINGHAM,
Merchant, Norfolk.

Some explanation of mercantile language used in relation to Bills of Exchange seems necessary, that the learner may have a clear idea of the questions, which will be given for practice.

When a merchant in the United States draws on his banker in London, his draft is styled " Bill on London" or " United States on London;" and if he sells his Bill at more than a dollar for 54 d. sterling, the exchange is said to be above par; and if he sells at less than a dollar for 54 d. sterling, below par: if a merchant in London draws on his banker in the United States, his draft is styled "London on United States;" and if he sells his Bill at more than 54 d. sterling for the dollar, the exchange is said to be above par; and if he sells at less than 54 d. sterling for the dollar, below par.

If the merchant in London draws on his banker in Paris, it is "London on Paris," or "London on France." If the merchant in Charleston, S. C. draws on his banker in New York, it is "Charleston on New York." &c.

GREAT BRITAIN.

In Great Britain, accounts are kept in pounds, shillings, pence, and farthings, *Sterling*.

The par value of the United States dollar is 4 s. 6 d. sterling; therefore, the dollar is equal to $\frac{9}{40}$ of a pound sterling. Hence, any sum of sterling money, (the shillings and pence, if any, being expressed in a decimal,) may be reduced to Federal money, by multiplying by 40 and dividing by 9: and, any sum in Federal money may be reduced to sterling money, by multiplying by 9 and dividing by 40. Or, if sterling money be increased by $\frac{1}{3}$ of itself, the sum expresses the same value in the old currency of New England: and, conversely, if the old currency of New England, be decreased by $\frac{1}{4}$ of itself, the result is the expression in sterling money.

1. United States on London. Reduce £784 14 s. 10½ d. sterling to Federal money, at par.

2. London on United States. Reduce 3487 dollars 75 cents to sterling, at par.

3. United States on London. Reduce £2006 11s. sterling to Federal money; exchange at 4 per cent. below par.

4. London on United States. Reduce 4287 dollars 50 cents to sterling; exchange at 4 per cent. above par.

5. London on United States. Reduce 3646 dollars 50 cents to sterling; exchange at 2 per cent. below par.

6. United States on London: Reduce £4109 11s. 10d. sterling to United States currency; exchange at 7 per cent. above par.

7. United States on London. Reduce £5129 15s. 6d. sterling to Federal money; exchange at 5 per cent. above par.

The law assimilating the currency of Ireland to that of England, took effect in January 1826. All invoices, contracts, &c. are considered there, in law, British currency, unless otherwise expressed.

8. United States on Dublin (Ireland). Reduce £1834 2s. 10½d. sterling to Federal money; exchange at 4 per cent. above par.

FRANCE.

Accounts were kept in France previous to 1795, according to the old system, in livres, sous, and deniers.

12 deniers = 1 sol or sou;
20 sous   = 1 livre;
6 livres   = 1 ecu or crown, silver.

By the new system, accounts are kept in francs, decimes, and centimes.

10 centimes = 1 decime.
10 decimes = 1 franc.

The value of the franc is 18⅗ cents in Federal money.

80 francs   = 81 livres.

9. United States on France. Reduce 7232 francs 38 centimes to Federal money; exchange at 1 dollar for 5 francs 30 centimes.

10. France on United States. Reduce 4093 dollars 80 cents to money of France; exchange at 5 francs 30 centimes for the dollar.

11. France on United States. Reduce 1834 dollars 65 cents to French currency; exchange at 5 francs 40 centimes for a dollar.

12. United States on France. Reduce 20828 francs 67 centimes to Federal money; exchange at 1 dollar for 5 francs 38 centimes.

13. United States on France. Reduce 12893 francs 27 centimes to Federal money; exchange at 1 dollar for 5 francs 33 centimes.

HAMBURGH.

Accounts are kept here in *marks, schillings* or *sols,* and *pfenirgs, Lubs.*

12 pfenirgs =1 sol or schilling, Lubs;

16 schillings=1 mark, Lubs;

3 marks   =1 reichsthaler or rix dollar specie.

Accounts are also kept, particularly in exchanges, in *pounds, shillings,* and *pence, Flemish.*

12 pence or grotes=1 shilling.

20 shillings   =1 pound, Flemish.

The word *Lubs* originally meant money of Lubeck, which is the same with that of Hamburgh, and the term is intended to distinguish this money from the Flemish denominations, and also from the money of Denmark and other neighboring places.

The mark Lubs is worth $2\frac{2}{3}$ shillings Flemish, or 32 grotes; consequently the sol Lubs is 2 grotes Flemish, and the shilling Flemish 6 schillings Lubs.

Banco or bank money, in which exchanges are reckoned, and currency, are the two principal kinds of money.

Banco consists of the sums of money deposited by merchants and others in the bank, and inscribed in its books; which sums are not commonly drawn out, but are transferred from one person to another in payment of a debt or contract.

Current money, or currency, consists of the common coins of the city, in which expenses are mostly paid.

The bank money is more valuable than currency, and bears a premium varying from 18 to 25 per cent. This premium is called the agio.   For instance, when the agio is 20 per cent. 100 marks banco are valued at 120 marks currency.

The mark banco is valued in the United States at 33⅓ cents.

14. United States on Hamburgh.   Reduce 1148 marks, 5 schillings, 4 pfenirgs banco to Federal money; exchange at 33 cents per mark banco.

15. Hamburgh on United States.   Reduce 1245 dollars 75 cents to money of Hamburgh; exchange at 3 marks banco per dollar.

16. United States on Hamburgh.   Reduce 6194 marks 12 schillings banco to Federal money; exchange at 34 cents per mark banco,

17. United States on Hamburgh.   Reduce 8246 marks 8 schillings banco to Federal money; exchange at 35 cents per mark banco.

18. Hamburgh on United States.   Reduce 757 dollars 90 cents to money of Hamburgh; exchange at 1 mark banco for 33 cents.

AMSTERDAM AND ANTWERP.

In these places accounts were formerly kept in *florins*, *stivers*, and *pennings*; or in *pounds*, *shillings*, and *pence*, *Flemish*.

16 pennings = 1 stiver,
20 stivers = 1 florin or guilder.

In Flemish, 12 grotes or pence, or 6 stivers = 1 shilling,
20 shillings, or 6 florins = 1 pound;
2½ florins, or 50 stivers = 1 rix dollar.

By the new system, adopted in 1815, accounts are kept throughout the kingdom of the Netherlands in florins or guilders, and cents.

100 cents = 1 florin or guilder.

The par value of the florin, in Federal currency, is 40 cents.

19. United States on Amsterdam.   Reduce 13790 florins 15 stivers to Federal money; exchange at 36 cents per florin.

20. United States on Antwerp. Reduce 6281 florins 88 cents to Federal money; exchange at 40 cents per florin.

21. Amsterdam on United States. Reduce 2482 dollars 33½ cents to Dutch money; exchange at 36 cents per florin.

22. Antwerp on United States. Reduce 3436 dollars 72 cents to Dutch money; exchange at 38 cents per florin.

23. United States on Antwerp. Reduce 7294 florins 50 cents to Federal money; exchange at 42 cents per florin.

24. United States on Amsterdam. Reduce 10148 florins to Federal money; exchange at 41 cents per florin.

PORTUGAL.

In Portugal, accounts are kept in milrees and rees; and also in *old crusados*.

    1000 rees = 1 milree.

    400 rees = 1 old crusado or crusado of exchange.

    480 rees = 1 new crusado.

There are three sorts of money used in Portugal; viz. effective money, i. e. specie; paper money, which is at a discount; and legal money, consisting of half specie and half paper.

The value of the milree in Federal money is 1 dollar 24 cents.

25. United States on Lisbon. Reduce 964 milrees 475 rees to Federal money; exchange at 1 dollar 24 cents per milree.

26. Lisbon on United States. Reduce 1274 dollars 66 cents to money of Portugal; exchange at 1 dollar 25 cents per milree.

27. United States on Portugal. Reduce 1248 milrees 645 rees to Federal money; exchange at 1 dollar 26 cents per milree.

28. United States on Portugal. Reduce 1846 milrees 500 rees to Federal money; exchange at 1 dollar 23 cents per milree.

The exchanges of Brazil, in South America, are similar to those of Portugal; there is, however, a difference in the value of their moneys; that of Portugal is half specie and half paper, called legal money, and that of Brazil is effective.

SPAIN.

The most general mode of keeping accounts in Spain is in *reals*, of 34 maravedis; but there are nine different *reals*, each divided into 34 maravedis, but differing in value. Four of these *reals* are of general application, and five of local use.

The four principal moneys of Spain are the *real vellon*, the *real of old plate*, the *real of new plate*, and the *real of Mexican plate;* and in order to obtain a distinct view of them, it may be proper to make the *real vellon* the basis of all the rest.

The *real vellon* is the twentieth part of the hard dollar, (peso duro), universally known by the name of the Spanish dollar, which is the same in value with the dollar of the United States.

The division of the real vellon is into *quartos, ochavos,* and *maravedis*.

2 maravedis = 1 ochavo,
2 ochavos = 1 quarto,
8 quartos, or 34 maravedis = 1 real vellon;

but maravedis are commonly used to express any fraction of a real: thus, we say 1 real 33 maravedis.

The real of old plate is better than the real vellon, in the proportion of 32 to 17. Thus 17 maravedis of old plate are equal to 32 maravedis vellon; the quartos and ochavos are in the same proportion. The real of old plate is not a coin; it is the most general money of exchange. 8 of these reals make the *piastre*, which is also called the *dollar of exchange*. $10\frac{5}{8}$ of these reals are equal to the hard dollar. When plate only is mentioned, old plate is understood.

The real of new plate is double the real vellon; therefore 34 maravedis of new plate are equal to 68 of vellon;

quartos and ochavos in proportion. This real is a coin, but not a money of account in any general way; it is the tenth part of the hard dollar, and is estimated in the United States at 10 cents, and the real vellon at 5 cents.

The real of Mexican plate is divided into *halves* and *quarters*, called *medio* and *quartillo*. It is the eighth part of the hard dollar, and is the chief money of account in Spanish America, where it is divided into sixteenths.

The *doubloon de plata*, or *doubloon of exchange* is four times the value of the piastre, or dollar of exchange.

The *ducado de plata*, or *ducat of exchange* is worth 11 reals 1 maravedi old plate, or 20 reals 25 $\frac{1}{17}$ maravedis vellon.

29. United States on Spain. Reduce 3148 dollars (of exchange) 6 reals 32 maravedis plate to Federal money, exchange at 67 cents per piastre ?

30. Spain on United States. Reduce 1821 dollars 60 cents to Spanish money; exchange at 68 cents per dollar of exchange.

31. United States on Spain. Reduce 1286 dollars (of exchange) 7 reals 17 maravedis plate to Federal money; exchange at 64 cents per dollar of exchange.

32. United States on Spain. Reduce 2136 doubloons of exchange to Federal money; exchange at 68 cents per dollar (of exchange).

33. United States on Buenos Ayres. Reduce 4680 rials of Mexican plate to Federal money; exchange at 12 cents per rial.

SWEDEN.

In Sweden, accounts are kept in rix dollars specie, skillings, and rundstycken or ore.

12 rundstycken or ore = 1 skilling;
48 skillings = 1 rix dollar specie.

The Swedish dollar agrees in value with the dollar of the United States.

34. United States on Sweden. Reduce 3955 rix dollars 24 skillings to Federal money; exchange at 1 dollar 2 cents per rix dollar.

35. Sweden on the United States. Reduce 1344 dollars 87 cents Federal money to Swedish money; exchange at 1 rix dollar for 1 dollar 2 cents.

36. United States on Sweden. Reduce 2481 rix dollars 36 skillings to Federal money; exchange at 1 dollar per rix dollar.

37. Sweden on the United States. Reduce 819 dollars 87½ cents Federal money to Swedish currency; exchange at 1 rix dollar per dollar.

38. United States on Sweden. Reduce 1234 rix dollars 12¼ skillings to Federal money; exchange at 98 cents per rix dollar.

39. United States on Sweden. Reduce 1126 rix dollars 42 skillings to Federal money; exchange at 1 dollar 3 cents per rix dollar.

RUSSIA.

In Russia accounts are kept in roubles and copecks. The rouble is also divided into 10 grieven.

10 copecks = 1 grieve or grievener,
10 grieven or 100 copecks, = 1 rouble.

The silver rouble is estimated in the United States at 75 cents; but the commercial business of Russia is carried on, in a paper currency much inferior to that of specie. The variable agio of the paper, substituted for the silver rouble, makes exchanges with Russia extremely fluctuating, as the paper rouble improves or declines in value.

40. United States on Russia. Reduce 4182 roubles 64 copecks to Federal money; exchange at 25 cents per rouble.

41. Russia on the United States. Reduce 2614 dollars 15 cents to Russian money; exchange at 1 rouble for 25 cents.

42. United States on Russia. Reduce 5416 roubles 50 copecks to Federal money; exchange at 28 cents per rouble.

43. Russia on United States. Reduce $3148.56 to Russian currency; exchange at 1 rouble for 30 cents.

44. United States on Russia.   Reduce 8672 roubles 75 copecks to Federal money; exchange at 32 cents per rouble.

PRUSSIA.

In Prussia accounts are generally kept in thalers or rix dollars, good grosehen, and pfenings.

12 pfenings   = 1 good grosehen,
24 good grosehen = 1 rix dollar.

The Prussian rix dollar is in value $\frac{2}{3}$ of the dollar of the United States.

45. United States on Prussia.   Reduce 4162 rix dollars 18 good grosehen to Federal money; exchange at 66 cents per rix dollar.

46. Prussia on United States.   Reduce 3148 dollars 32 cents to Prussian money; exchange at 1 rix dollar for 64 cents.

47. United States on Prussia.   Reduce 1428 rix dollars 14 good grosehen to Federal money; exchange at 67 cents per rix dollar.

48. Prussia on United States.   Reduce 2136 dollars Federal money to Prussian money; exchange at 1 rix dollar for 48 cents.

DENMARK.

In 1813 a new monetary system was established in Denmark, in which system the rigsbank dollar is the money unit.   The denominations of money are the same as in the old, or Hamburgh system, but of only half the value.

12 pfenings = 1 skilling,
16 skillings = 1 mark,
6 marks   = 1 rigsbank dollar.

The Danish rigsbank dollar is equal to 50 cents in the United States.

49. United States on Denmark.   Reduce 3214 rigsbank dollars 4 marks 8 skillings to Federal money; exchange at 50 cents per rigsbank dollar.

50. Denmark on the United States. Reduce 2082 dollars 35 cents Federal money to Danish money; exchange at 2 rigsbank dollars per dollar.

51. United States on Denmark. Reduce 1968 rigsbank dollars 5 marks 12 skillings to Federal money; exchange at 48 cents per rigsbank dollar.

52. Denmark on United States. Reduce 3007 dollars Federal money to Danish money; exchange at 2 rigsbank dollars 2 skillings per dollar.

NAPLES.

In Naples accounts are kept in ducati, carlini, and grani. The ducat is the money unit, and is divided into 10 carlins, each of 10 grains, and, by the public banks, into 5 tarins of 20 grains each, making the ducat always 100 grains.

$$10 \text{ grani} = 1 \text{ carlino,}$$
$$10 \text{ carlini} = 1 \text{ ducato.}$$

The value of the silver ducat, in Federal money, is 80 cents.

53. United States on Naples. Reduce 4022 ducati 8 carlini to Federal money; exchange at 80 cents per ducat.

54. Naples on United States. Reduce 1835 dollars 73 cents Federal money to Neapolitan money; exchange at 1 ducato for 78 cents.

55. United States on Naples. Reduce 3508 ducats 5 carlini to Federal money; at 82 cents per ducat.

56. Naples on the United States. Reduce 1817 dollars 54 cents Federal money to Neapolitan money; exchange at 1 ducat for 76 cents.

SICILY.

In Sicily accounts are kept in oncie, tari, and grani.

$$20 \text{ grani} = 1 \text{ taro,}$$
$$30 \text{ tari} = 1 \text{ oncia.}$$

Accounts are also kept in scudi, tari, and grani.

$$12 \text{ tari} = 1 \text{ scudo, or Sicilian crown,}$$
$$5 \text{ scudi} = 2 \text{ oncie.}$$

It must be observed, that the denominations of Sicilian money have but half the value of the same denominations in Naples. The Sicilian oncia, for instance, passes in Naples for only 15 tari, the Sicilian scudo for 6 tari, and other denominations in the same proportion.

The scudo is equal to 96 cents, and the oncia to $2.40 Federal money.

57. United States on Sicily. Reduce 1214 oncie 20 tari 10 grani to Federal money; exchange at 8 cents per taro.

58. Sicily on United States. Reduce 1457 dollars 62 cents Federal money to Sicilian money; exchange at 1 taro for 8 cents.

59. United States on Sicily. Reduce 3010 scudi 9 tari 15 grani to Federal money; exchange at 96 cents per scudo.

60. Sicily on United States. Reduce 983 dollars 44 cents to Sicilian money; enchange at 1 scudo for 95 cents.

LEGHORN.

In Leghorn accounts are kept in pezze, soldi, and denari di pezza.

12 denari di pezza = 1 soldo,
20 soldi di pezza = 1 pezza of 8 reals.

The value of the pezza is 90 cents, Federal money.

61. United States on Leghorn. Reduce 2146 pezze 16 soldi 8 denari to Federal money; exchange at 92 cents per pezza of 8 reals.

62. Leghorn on United States. Reduce 1620 dollars 45 cents to money of Leghorn; exchange at 1 pezza of 8 reals for 90 cents.

63. United States on Leghorn. Reduce 3293 pezze 13 soldi 4 denari to Federal money; exchange at 93 cents per pezza of 8 reals.

64. Leghorn on United States. Reduce 1214 dollars 68 cents to money of Leghorn; exchange at 1 pezza of 8 reals for 91 cents.

GENOA.

In Genoa accounts are kept in lire, soldi, and denari di lira; or in pezze, soldi, and denari di pezza; all in money fuori banco, or current money.

12 denari = 1 soldo,
20 soldi = 1 lira;
12 denari = 1 soldo,
20 soldi = 1 pezza;

The value of the pezza is to that of the lira as 4 to 23; that is, 4 pezze = 23 lire; therefore 4 soldi di pezza = 23 soldi di lira; 4 denari di pezza = 23 denari di lira.

The par value of the lira is $15\frac{1}{2}$ cents, that of the pezza 89 cents, U. S.

65. United States on Genoa. Reduce 5254 lire 16 soldi 3 denari to Federal money; exchange at 16 cents per lira.

66. Genoa on the United States. Reduce 1532 dollars 30 cents to money of Genoa; exchange at 1 lira for 15 cents.

67. United States on Genoa. Reduce 8792 lira to Federal money; exchange at $16\frac{1}{2}$ per lira;

68. Genoa on the United States. Reduce 2000 dollars to money of Genoa; exchange at 1 lira for $15\frac{1}{2}$ cents.

VENICE.

In Venice accounts were formerly kept, and exchanges computed in ducats, lire, soldi, and denari, moneta piccola.

12 denari   = 1 soldo,
20 soldi   = 1 lira piccola,
$6\frac{1}{4}$ lire piccole = 1 ducat current,
8 lire piccole = 1 ducat effective.

The par of the lira piccola, in Federal money, is $9\frac{3}{5}$ cents.

Accounts are now kept, and exchanges computed in lire Italiane and centimes.

100 centesimi = 1 lira Italiana.

The common estimate of this money is, that 100 lire

piccole are equal to 51⅙ lire Italiane. The lira Italiana is of the same value with the French franc.

. 69. United States on Venice. Reduce 14642 lire 4 soldi 8 denari piccoli to Federal money; exchange at 9 cents per lira piccola.

70. Venice on the United States. Reduce 814 dollars 55 cents to Venetian money; exchange at 1 lira piccola for 8 cents.

We have given the two preceding examples in the old currency, for the sake of practice, although it has generally gone out of use.

71. United States on Venice. Reduce 6784 lire Italiane 40 centimes to Federal money; exchange at 18½ cents per lira Italiana.

72. Venice on the United States. Reduce 1817 dollars 82 cents to Venetian money; exchange at 1 lira Italiana for 18 cents.

73. United States on Venice. Reduce 5236 lire Italiane to Federal money; exchange at 18¾ cents per lira Italiana.

TRIESTE.

In Trieste accounts are kept and exchanges computed in florins and creutzers; or in rix dollars and creutzers.

$$4 \text{ pfenings} = 1 \text{ creutzer,}$$
$$60 \text{ creutzers} = 1 \text{ florin or gulden,}$$

1½ florin, or 90 creutzers = 1 rix dollar of account.

The rix dollar specie is equal to 2 florins.

The par of the florin is 48 cents, in Federal money, which makes the dollar of account 72 cents, and the specie dollar 96 cents.

74. United States on Trieste. Reduce 2846 florins 25 creutzers to Federal money; exchange at 48 cents per florin.

75. Trieste on United States. Reduce 1637 dollars 48 cents to money of Trieste; at 1 florin for 47 cents.

76. United States on Trieste. Reduce 2055 rix dollars, 25 creutzers to Federal money; exchange at 72 cents per rix dollar.

77. Trieste on United States. Reduce 1738 dollars 38 cents to money of Trieste; exchange at 1 rix dollar for 70 cents.

ROME.

In Rome accounts are kept, by the old system, in scudi, paoli, and bajocchi; quattrini and mezzi quattrini are also sometimes reckoned.

$$2 \text{ mezzi quattrini} = 1 \text{ quattrino,}$$
$$5 \text{ quattrini} \qquad = 1 \text{ bajoccho,}$$
$$10 \text{ bajocchi} \qquad = 1 \text{ paolo,}$$

10 paoli, or 100 bajocchi = 1 scudo, or Roman crown.

The Roman crown is equal to the Federal dollar.

The scudo di stampa d'oro, or gold crown, is equal to $1.53.

When the exchange between the United States and Rome is at par, no reduction is required; for any number of scudi and bajocchi are equal to the same number of dollars and cents, and the reverse; for instance, 125 scudi 75 bajocchi are equal to 125 dollars 75 cents.

78. Rome on the United States. Reduce 1871 dollars 19 cents to Roman money; exchange at 1 scudo 2 bajocchi per dollar.

79. United States on Rome. Reduce 2070 scudi 50 bajocchi to Federal money; exchange at 101 cents per scudo.

In 1809, the French moneys of account were introduced into Rome. The scudo was reckoned at 5 francs 35 centimes; the franc, therefore, was valued at 18 bajocchi 3.45 quattrini.

MALTA.

Acounts are kept in this island in scudi, tari, and grani,

$$20 \text{ grani} = 1 \text{ taro,}$$
$$12 \text{ tari} \quad = 1 \text{ scudo.}$$

The taro is likewise divided into 2 carlini, and a carlino into 60 piccioli. The pezza, or dollar of exchange, is equal to $2\frac{1}{2}$ scudi.

The par value of the Maltese scudo is 40 cents in Federal money.

The coins in circulation are chiefly Spanish dollars and doubloons, and Sicilian dollars and ounces. They are valued each at a certain rate, as follows, on which a variable agio is charged.

> Spanish dollar   = 30 tari 10 grani.
> Spanish doubloon = 38 scudi 9 tari.
> Sicilian dollar   = 30 tari.
> Sicilian ounce   = 6 scudi 3 tari.

80. United States on Malta. Reduce 1108 Maltese scudi 9 tari to Federal money; exchange at 40 cents per scudo.

81. Malta on the United States. Reduce 874 dollars 76 cents to Maltese money; at 1 scudo for 38 cents.

82. United States on Malta. Reduce 3964 Maltese scudi 6 tari to Federal money; at 41 cents per scudo.

83. Malta on the United States. Reduce 674 dollars 60 cents to Maltese money; at 1 scudo for 40 cents.

SMYRNA.

Accounts are kept here in piastres or gooroosh. The piastre, also called the Turkish dollar, is divided sometimes into 12 temins, sometimes into 40 paras or medini; but the usual division is into aspers, the number of which varies. Thus, the English and Swedes divide the piastre into 80 aspers; the Dutch, French, and Venetians into 100 aspers; the Turks, Greeks, Persians, and Armenians into 120 aspers. An asper is a third part of a para.

Bills of exchange are often drawn on Smyrna in foreign coin, particularly in Spanish dollars, which are always to be had there; but, if drawn in a coin not in current use, the exchange of the day is established to make the payment.

The Turkish coins, owing to the frequent deterioration of them by the government, have been declining in their intrinsic worth for many years, and have no standard value. Foreign exchanges are conducted entirely according to the price of the day.

84. United States on Smyrna. Reduce 5318¾ piastres of Turkey to Federal money; exchange at 20 cents per piastre.

85. Smyrna on the United States. Reduce 912 dollars 27 cents to Turkish money; exchange at 1 piastre for 21 cents.

86. United States on Smyrna. Reduce 7161½ Turkish piastres to Federal money; exchange at 22 cents per piastre.

87. Smyrna on United States. Reduce 1128 dollars 22 cents to money of Smyrna; exchange at 1 piastre for 20½ cents.

EAST INDIES.

Before European colonies were established in the East Indies, particularly while the power of the Moguls prevailed in Hindostan, the monetary system was very simple. There was current throughout these vast dominions one principal coin of silver, denominated the *sicca rupee*. It was of a certain weight called the *sicca*. The *sicca* was used also as a standard for weighing other articles.

The British possessions in the East Indies are divided into three presidencies, viz. Bengal, Bombay, and Madras. The monetary systems in these presidencies are different from each other.

CALCUTTA IN BENGAL.

Accounts are commonly kept here in *current rupees, annas,* and *pice.*

12 pice = 1 anna,
16 annas = 1 rupee, currency.

The East India Company, however, keep their accounts in sicca rupees, similarly divided, which bear a *batta* or *premium* of 16 per cent. above current rupees.

The current rupee of Calcutta is $44\frac{4}{10}$ cents, and the sicca rupee $51\frac{4}{10}$ cents, in Federal money.

A Lac of rupees is 100000, and a Crore of rupees is 100 Lacs, or 10 millions of rupees.

88. United States on Calcutta. Reduce 17438 rupees 12 annas, currency of Calcutta, to Federal money; exchange at 48 cents per rupee.

89. Calcutta on the United States. Reduce 6913 dollars .25 cents to money of Calcutta; exchange at 2 sicca rupees per dollar.

90. United States on Calcutta. Reduce 46173 current rupees 9 annas to Federal money; exchange at 46 cents per rupee.

91. Calcutta on United States. Reduce 28953 dollars 63 cents to current money of Calcutta; exchange at 1 rupee for 44 cents.

92. United States on Calcutta. Reduce a Lac of sicca rupees to Federal money; exchange at 53 cents per sicca rupee.

BOMBAY.

In the presidency of Bombay, accounts are kept in rupees, quarters, and reas.

$$100 \text{ reas} = 1 \text{ quarter,}$$
$$4 \text{ quarters} = 1 \text{ rupee.}$$

The current value of the Bombay rupee is equal to 50 cents in Federal money.

93. United States on Bombay. Reduce 10137 rupees 2 quarters 50 reas to Federal money; exchange at 50 cents per rupee.

94. Bombay on the United States. Reduce 6210 dollars 48 cents to money of Bombay; exchange at 48 cents per rupee.

95. United States on Bombay. Reduce 8413 rupees of Bombay to Federal money; at 49 cents per rupee.

MADRAS.

In the presidency of Madras, there are different monetary systems, which may be distinguished under the heads of the old system and the new.

According to the old system, accounts are kept in *star pagodas, fanams,* and *cash.*

80 cash   = 1 fanam,
42 fanams = 1 pagoda.

The current value of the star pagoda is $ 1.80 Federal.

By the new system, the silver rupee of Madras is made the standard coin, and money of account in this presidency.

The current value of the silver rupee of Madras is $44\frac{4}{10}$ cents. It is divided into halves, quarters, eighths, and sixteenths. The sixteenth is the anna.

96. United States on Madras. Reduce 7218¾ rupees of Madras to Federal money; at 46 cents per rupee.

97. Madras on United States. Reduce 2684 dollars 88 cents to money of Madras; at 44 cents per rupee.

98. United States on Madras. Reduce 5367 rupees of Madras to Federal money; at 45 cents per rupee.

CANTON IN CHINA.

In China accounts are kept in *tales, mace, candarines*, and *cash.*

10 cash   = 1 candarine,
10 candarines = 1 mace,
10 mace   = 1 tale.

The tale is reckoned at $ 1.48 in Federal money.

99. United States on Canton. Reduce 12144 tales 5 mace to Federal money; exchange at 1 dollar 48 cents per tale.

100. Canton on the United States. Reduce 8754 dollars 89 cents to money of Canton; exchange at 1 tale per 146 cents.

101. United States on Canton. Reduce 16235 tale to Federal money; exchange at 149 cents per tale.

JAPAN.

In the empire of Japan, which consists of several islands to the east of Asia, accounts are kept in *tales, mace,* and *candarines.*

10 candarines = 1 mace,
10 mace   = 1 tale.

The Japanese tale is reckoned at 75 cts, Fed. money.

102. United States on Japan. Reduce 3714 Japanese tales 8 mace 8 candarines to Federal money; exchange at 75 cents per tale.

103. Japan on United States. Reduce 696 dollars 54 cents to money of Japan; exchange at 1 tale per 75 cents.

104. United States on Japan. Reduce 2468 tales 5 mace, money of Japan, to Federal money; exchange at 76 cents per tale.

SUMATRA.

This island is chiefly in possession of the natives; but the English have a small settlement at Bencoolen.

At Bencoolen accounts are kept in *dollars, soocoos,* and *satellers.*

$$8 \text{ satellers} = 1 \text{ soocoo,}$$
$$4 \text{ soocoos} = 1 \text{ dollar.}$$

This dollar is reckoned at $1.10 in Federal money, and is sometimes called a rial.

105. United States on Bencoolen. Reduce 1947 Bencoolen dollars 3 soocoos 4 satellers to Federal money; exchange at 110 cents per dollar of Bencoolen.

106. Bencoolen on United States. Reduce $2379.51 Federal money to money of Bencoolen; exchange at 1 dollar Bencoolen for 103 cents.

ACHEEN. (*In the island of Sumatra*).

In Acheen accounts are kept in *tales, pardows, mace,* and *copangs.*

$$4 \text{ copangs} = 1 \text{ mace,}$$
$$4 \text{ mace} = 1 \text{ pardow,}$$
$$4 \text{ pardows} = 1 \text{ tale.}$$

The mace is a small gold coin worth about 26 cents Federal money, which makes the tale $4.16.

107. United States on Acheen. Reduce 1432 tales, 3 pardows 2 mace to Federal money; exchange at 416 cents per tale.

108. Acheen on United States. Reduce 3620 dollars 96½ cents to money of Acheen; at 1 tale for 412 cents.

JAVA.

In Batavia, the capital of this island, the florin or guilder of the Netherlands is the monetary unit; but instead of the decimal divisions, it is here sometimes divided into *schillings, dubbels, stivers,* and *doits.*

5 doits   = 1 stiver,
2 stivers = 1 dubbel,
3 dubbels = 1 schilling,
4 schilling = 1 florin or guilder.

The florin of Java, as the florin of the Netherlands, is equal to 40 cents Federal money.

109. United States on Batavia. Reduce 11841 florins 3 schillings 2 dubbels to Federal money; exchange at 40 cents per florin.

110. Batavia on the United States. Reduce $13746. 69 to money of Batavia; exchange at 1 guilder for 38 cents.

111. United States on Batavia. Reduce 42328 guilders 50 centimes to Federal money; exchange at 42 cents per guilder.

MANILLA. (*In the island of Luzon*).

In Manilla, the capital of the Spanish East India possessions, accounts are kept in *Spanish dollars* or *pesos, reals,* and *maravedis.*

34 maravedis = 1 real,
8 reals   = 1 dollar.

112. United States on Manilla. Reduce 6341 dollars 6 reals 17 maravedis to Federal money; exchange at 101 cents per Spanish dollar.

113. Manilla on United States. Reduce $5274.55 to money of Manilla; exchange at 1 Spanish dollar per dollar.

COLOMBO. (*In the island of Ceylon*).

In Colombo, accounts are kept in *rix dollars, fanams,* and *pice.*

4 pice   = 1 fanam,
12 fanams = 1 rix dollar.

The current value of this rix dollar is 40 cts. F. money.

114. United States on Colombo. Reduce 7328 rix dollars 9 fanams to Federal money; exchange at 40 cents per rix dollar.

115. Colombo on United States. Reduce $1426.71 Federal money to money of Colombo; exchange at 1 rix dollar per 38 cents.

MAURITIUS, (*Isle of France.*)

In Mauritius, accounts are kept in two different ways, viz. in *dollars* of 100 cents, which is the mode adopted in public or government accounts; and in *dollars, livres,* and *sols,* which method is mostly used by merchants.

20 sols   = 1 livre.
10 livres = 1 dollar.

These are called colonial livres, and are 10 cents each.

116. United States on Mauritius. Reduce 4132 dollars 7 livres 10 sols to Federal money; exchange at 1 dollar per dollar of Mauritius.

117. Mauritius on United States. Reduce $7547.47, Federal money, to money of Mauritius; exchange at 98 cents per dollar of Mauritius.

ARBITRATION OF EXCHANGE.

ARBITRATION OF EXCHANGE is a comparison of the courses of exchange between different countries, in order to ascertain the most advantageous course of drawing or remitting bills. It is distinguished into simple and compound arbitration.

Simple Arbitration is a comparison between the exchanges of two places through a third; that is, it is finding such a rate of exchange between two places, as shall be in proportion to the rates quoted between each of them and a third place. The exchange thus determined is called the arbitrated price.

If, for example, the course of exchange between London and Paris is 24 francs for 1 pound sterling, and between Paris and Amsterdam 54 pence Flemish for 3 francs, the arbitrated price between London and Amsterdam through Paris, is 36 shillings Flemish for 1 pound sterling; for, as 3 fr. : 24 fr. = 54 d. : 36 s. Flem.

Suppose the arbitrated price to be, as before stated, 36 s. Flemish for £1 sterling; and suppose the direct course between London and Amsterdam to be 37 s. Flemish; then London, by drawing directly on Amsterdam, must give 37 s. Flemish for £1 sterling; whereas, by drawing through Paris, he will give only 36 s. Flemish, for £1 sterling. It is therefore the interest of London to draw indirectly on Amsterdam through Paris.

On the contrary, if London remits directly to Amsterdam, London will receive 37 s. Flemish for £1 sterling; but, by remitting through Paris, London will receive only 36 s. Flemish. It is the interest of London, therefore, to remit directly to Amsterdam.

118. If the exchange of London with Genoa is 47 d. sterling per pezza, and that of Amsterdam with Genoa 86 grotes Flemish per pezza, what is the proportional or arbitrated exchange between London and Amsterdam through Genoa? that is, how many shillings and grotes Flemish are equal to £1 sterling?

Since 47 d. sterling is equal to 1 pezza, and this pezza is equal to 86 grotes Flemish, the question may be stated thus; 47 d. sterling : 240 d. sterling = 86 grotes Flemish: *Ans.* which is 36 s. $7\frac{7}{17}$ grotes Fl. By the Chain Rule, (See Art. xxvi), the statement is as follows.

$$1 \text{ pound sterling.}$$

1 pound sterling = 240 pence.
47 pence   = 1 pezza.
1 pezza   = 86 grotes Fl.
12 grotes   = 1 shilling Fl.

The product of the consequents being divided by the product of the antecedents, will give 36 sh. $7\frac{7}{17}$ grotes Fl. for the answer.

119. If the exchange on London with Hamburgh is 34 shillings 2 grotes Flemish banco for £1 sterling, and

that of Amsterdam with Hamburgh 33⅓ stivers per rix dollar of 2 marks, what is the arbitrated exchange between London and Amsterdam through Hamburgh?

Since 2 marks are 64 grotes Flemish and 33⅓ stivers are 66¾ grotes Flemish, the question may be stated thus, 64 grotes Fl.: 66¾ grotes Fl. = 34 s. 2 grotes Fl.: Ans. By the Chain Rule, the statement is as follows,

> 1 pound sterling.
> 1 pound sterling = 34 s. grotes Flem.
> 8 s. Flem.   = 3 marks.
> 2 marks   = 33⅓ stivers.
> 6 stivers   = 1 s. Flemish.

120. If the exchange of London on Leghorn is 51½ d. sterling per pezza, and that of Amsterdam on Leghorn 92¾ grotes Flemish per pezza, what is the proportional exchange between London and Amsterdam through Leghorn?

121. If the exchange of London on Lisbon be 68 d. sterling per milree, and that of Amsterdam on Lisbon 48 grotes Flemish per old crusado, what is the arbitrated exchange between London and Amsterdam through Lisbon?

122. If the exchange of London on Madrid is 42 d. sterling per dollar of plate, and that of Amsterdam on Madrid 96 grotes Flemish per ducat of plate, what is the proportional exchange between London and Amsterdam through Madrid?

123. If the exchange of London on Paris is 24 francs per £1 sterling, and that of the United States on Paris 18½ cents per franc, what is the arbitrated or proportional exchange between London and the United States through Paris?

124. If the exchange of London on Amsterdam is 11 florins 16 stivers per £ sterling, and that of the United States on Amsterdam 38 cents per florin, what is the arbitrated exchange between the United States and London through Amsterdam?

125. If the exchange of the United States on Paris is 18 cents per franc, and that of Amsterdam on Paris 54 grotes Flemish for 3 francs, what is the proportional exchange between the United States and Amsterdam through Paris?

126. If the exchange of the United States on Lisbon is $1.24 per milree, and that of Paris on Lisbon 540 rees per ecu of 3 francs, what is the proportional exchange between the United States and Paris through Lisbon?

COMPOUND ARBITRATION.

Compound arbitration is a comparison between the exchanges of more than three places, to find the arbitrated price between the first place and the last, in order to determine on the most advantageous mode of negotiating bills.

127. Suppose the exchange between London and Amsterdam to be 35 shillings Flemish for £1 sterling; between Amsterdam and Lisbon, 42 pence Flemish per old crusado; and between Lisbon and Paris, 480 rees' per ecu of 3 francs; what is the arbitrated price between London and Paris?

First, 35 s. Fl. : 42d. Fl. = £1 sterling : A; which is 2 s. sterling.

Secondly, 1 old crusado : 480 rees = 2 s. sterling : A; which is 2 s. 4⅘ d. sterling.

Thirdly, 2 s. 4⅘ d, sterling : £1 sterling = 3 francs : A; which is 25 francs.

Hence the arbitrated price is 25 francs for £1 sterling.

But all such operations are best performed by the Chain Rule; thus,   1 pound sterling.

1 pound sterling =   35 shillings Flemish.
3½ shillings Fl. =   1 old crusado.
1 old crusado   = 400 rees.
480 rees   =   3 francs.

The product of the consequents divided by that of the antecedents gives 25 francs per £ sterling, as before.

128. Suppose a merchant in London has a sum of money to receive in Cadiz, the exchange being at 38 d. sterling per dollar of plate; but, instead of drawing directly on Cadiz, he draws on Amsterdam, ordering his agent there to draw on Paris, and Paris to draw on Cadiz; the exchange between London and Amsterdam being

at 35 shillings Flemish per pound sterling; between Amsterdam and Paris $53\frac{1}{2}$ grotes Flemish per ecu of 3 francs; and between Paris and Cadiz 15 francs 50 centimes per doubloon of plate. What is the arbitrated price between London and Cadiz?

$$\begin{aligned}
&\qquad\qquad\qquad\quad\text{1 dollar of plate.}\\
&\text{4 dollars of plate} \;=\; \text{1 doubloon of plate.}\\
&\text{1 doubloon of plate} = 15\tfrac{1}{2}\text{ francs.}\\
&\text{3 francs} \qquad\qquad = 53\tfrac{1}{2}\text{ grotes Fl.}\\
&\text{12 grotes Fl.} \qquad = \text{1 shilling Fl.}\\
&\text{35 shillings Fl.} \qquad =240\text{ pence sterling.}
\end{aligned}$$

The result is $39\frac{41}{64}$ d. sterling per dollar of plate. The circuitous operation is, therefore, the most advantageous, as London gets $39\frac{1}{2}$ d. nearly, instead of 38 d. for each dollar of plate.

129. London having a sum to receive in Lisbon, when the exchange is at 64 d. sterling per milree, draws on Lisbon, but remits his bill to Hamburgh to be negotiated, and directs the returns to be made to him in bills on Leghorn; the exchange between Hamburgh and Lisbon being 45 grotes Flemish per old crusado; between Hamburgh and Leghorn 85 grotes Flemish per pezza; and between London and Leghorn 52 d. sterling per pezza. What is the arbitrated price between London and Lisbon? and what does London give per milree by the circuitous exchange?

130. A merchant in London has a sum to pay in Petersburg, and another to receive in Genoa; but there being no regular exchange between these places, London draws on Hamburgh, and remits his bill to Petersburg, directing Hamburgh to draw on Genoa; the exchange between London and Genoa being $46\frac{1}{2}$ d. sterling per pezza; between Hamburgh and Genoa 81 grotes Flemish per pezza; and between Petersburg and Hamburgh 23 schillings Lubs per ruble. What is the exchange between London and Petersburg resulting from the operation? that is, how many pence sterling does London pay for the ruble?

131. A merchant in the United States has funds in Paris, and owes a sum of money in Hamburgh; he draws

on London, remits his bill to Hamburgh, and directs London to draw on Paris; the exchange between the United States and Paris being 18 cents per franc; between London and Paris 24 francs 25 centimes per £ sterling; and between Hamburgh and London 13½ marks banco per £ sterling. How many cents per mark banco does the American merchant pay by this course of exchange?

132. A merchant in the United States being indebted in London, remits bills on Paris to his correspondent in that city, and directs him to obtain bills of Paris on Lisbon and remit them to his creditor in London; the exchange between the United States and France being 18 cents per franc; between Paris and Lisbon 465 rees per ecu of 3 francs; and between London and Lisbon 63 d. sterling per milree. In this course of exchange, how many pence sterling are paid with one dollar of the United States?

In the preceding examples, no notice is taken of the expenses incident to exchange operations, such as commission, brokerage, interest, &c.; but in all transactions of business, it is necessary to make allowance for the difference of charges between direct and indirect exchanges, in order to decide on the preference of the one to the other.

FOREIGN COINS.

THE SILVER COINS of foreign countries, rendered current in the United States, by Act of Congress, are as follows. Spanish dollars and parts thereof, at 100 cents the dollar. Dollars of Mexico, Peru, Chili, and Central America, of not less weight than 415 grains each, and those restamped in Brazil of the like weight, and of not less fineness than 10 oz. 15 dwt. pure silver in the Troy pound, all at 100 cents the dollar. The Five-franc pieces of France, weighing 384 grains each, and of not less fineness than 10 oz. 16 dwt. pure silver in the Troy pound, at 93 cents the piece.

THE GOLD COINS of foreign countries, with their respective weights, and values, are stated in the following TABLE. Those of the countries printed in Italics are rendered current, by Act of Congress.

Names of Countries and Coins.	Weight. dwt. gr.		Value. $ cts m		
AUSTRIAN DOMINIONS.					
Souverein, ...	3	14	3	37	7
Double Ducat, ...	4	12	4	58	9
Hungarian, Ducat, . . .	2	5¾	2	29	6
BAVARIA					
Carolin, ...	6	5¼	4	95	7
Max d'or, or Maximilian, . .	4	4	3	31	8
Ducat, ...	2	5¾	2	27	5
BERNE.					
Ducat, double in proportion, . .	1	23	1	98	6
Pistole.	4	21	4	54	2
BRAZIL.					
Jahannes, half in proportion . .	18		17	06	4
Dobraon, ...	34	12	32	70	6
Dobra	18	6	17	30	1
Moidore, half in proportion, . .	6	22	6	55	7
Crusade, ...		16¼		63	5
BRUNSWICK.					
Pistole, double in proportion, .	4	21½	4	54	8
Ducat, ...	2	5¾	2	23	
COLOGNE.					
Ducat, ...	2	5¾	2	26	7
COLOMBIA.					
Doubloon	17	9	15	53	5
DENMARK.					
Ducat, Current, ...	2		1	81	2
Ducat, Specie, ...	2	5¾	2	26	7
Christian d'or, ...	4	7	4	02	1
EAST INDIA.					
Rupee, Bombay, 1818, . .	7	11	7	09	6
Rupee, Madras, 1818, . .	7	12	7	11	
Pagoda, Star, ...	2	4¾	1	79	8
ENGLAND.					
Guinea, half in proportion, . .	5	8½	5	07	5
Sovereign, half in proportion, .	5	2½	4	84	6

Seven Shilling Piece, . . .	1 19	1 69 8
FRANCE.		
Louis, coined before 1786, . .	5 5½	4 84 6
Double Louis, before 1786, . . .	10 11	9 69 7
Louis, coined since 1786, . .	4 22	4 57 6
Double Louis, since 1786, . .	9 20	9 15 3
Napoleon, or 20 francs, . . .	4 3½	3 85 1
Double Napoleon or 40 francs, .	8 7	7 70 2
Same as the new Louis Guinea, .	5	4 65 5
FRANKFORT ON THE MAIN.		
Ducat, ...	2 5¾	2 27 9
GENEVA.		
Pistole, old, ...	4 7¼	3 98 5
Pistole, new, ...	3 15¾	3 44 4
HAMBURG.		
Ducat, double in proportion, .	2 5¾	2 27 9
GENOA.		
Sequin, ...	2 5¾	2 30 2
HANOVER.		
Double George d'or, single in pro.	8 13	7 87 9
Ducat, ...	2 5¾	2 29 6
Gold Florin, double in pro'n, .	2 2	1 67
HOLLAND.		
Double Ryder, ...	12 21	12 20 5
Ryder, ...	6 9	6 04 3
Ducat, ...	2 5¾	2 27 5
Ten Guilder Piece, 5 do. in pro'n,	4 8	4 03 4
MALTA,		
Double Louis, ...	10 16	9 27 8
Louis, ...	5 8	4 65 2
Demi Louis, ...	2 16	2 33 6
MEXICO.		
Doubloons, shares in pro'n, . .	17 9	15 53 5
MILAN.		
Sequin, ...	2 5¾	2 29
Doppia or Pistole, ...	4 1½	3 80 7
Forty Livre Pieces, 1808, . .	8 8	7 74 2
NAPLES.		
Six Ducat Piece, 1783, . . .	5 16	5 24 9
Two do. or Sequin, 1762, . .	1 20¼	1 59 1
Three do. or Oncetta, 1818, .	2 10¼	2 49

NETHERLANDS.

Gold Lion, or 14 Florin Piece,	5	7¾	5	04	6
Ten Florin Piece, 1820, . . .	4	7¼	4	01	9

PARMA.

Quadruple Pistole, double in pro'n,	18	9	16	62	8
Pistole or Droppia, 1787, . .	4	14	4	19	4
Pistole or Droppia, 1796, . .	4	14	4	13	5
Maria Theresa, 1818, . . .	4	3½	3	86	1

PIEDMONT.

Pistole c'd since 1785, ½ in pro'n,	5	20	5	41	1
Sequin, half in proportion, . .	2	5	2	28	
Carlino, c'd since 1785, ½ in pro'n,	29	6	27	34	
Piece of 20 Francs, or Marengo,	4	3¼	3	56	4

POLAND.

Ducat, ...	2	5¾	2	27	5

PORTUGAL.

Dobraon, ...	34	12	32	70	8
Dobra, ...	18	6	17	30	1
Johannes, ...	18		17	06	4
Moidore, half in proportion, . .	6	22	6	55	7
Piece of 16 Testoons, 1600 Rees,	2	6	2	12	1
Old Crusado or 400 Rees, . .		15		58	8
New Crusado or 480 Rees, . .		16¼		63	5
Milree, coined in 1755, . .		19¾		73	

PRUSSIA.

Ducat, 1748, ...	2	5¾	2	27	9
Ducat, 1787, ...	2	5¾	2	26	7
Frederick, double, 1769, . .	8	14	7	97	5
Frederick, double, 1800, . .	8	14	7	95	1
Frederick, single, 1878, . . .	4	7	3	99	7
Frederick, single, 1800, . . .	4	7	3	97	5

ROME.

Sequin, coined since 1760, . .	2	4¼	2	25	1
Scudo of Republic, ...	17	0¼	15	81	1

RUSSIA.

Ducat, 1796, ...	2	6	2	29	7
Ducat, 1763, ...	2	5¾	2	26	7
Gold Ruble, 1756, ...	1	0½		96	7
Gold Ruble, 1799, ...		18¾		73	7
Gold Polten, 1777, ...		9		35	5
Imperial, 1801, ...	7	17¼	7	82	9

Half Imperial, 1801, ...	3	20½	3 91 8	
Half Imperial, 1818, ...	4	3½	3 93 3	
SARDINIA.				
Carlino, half in proportion, . .	10	7½	9 47 2	
SAXONY.				
Ducat, 1784, ...	2	5¾	2 26 7	
Ducat, 1797, ...	2	5¼	2 27 9	
Augustus, 1754, ...	4	6½	3 92 5	
Augustus, 1784, ...	4	6½	3 97 4	
SICILY.				
Ounce, 1751, ...	2	20½	2 50 4	
Double Ounce, 1758,. . . .	5	17	5 04 4	
SPAIN.				
Doubloons, 1772, double and sin-gle and shares in proportion, . .	17	8½	16 02 8	
Doubloon, ...	17	9	15 53 5	
Pistole, ...	4	8¼	3 88 4	
Coronilla, (Gold Dol.) or Vintern, 1801,	1	3	98 3	
SWEDEN.				
Ducat, ...	2	5	2 23 5	
SWITZERLAND.				
Pistole of Helvetic Republic, 1800,	4	21½	4 56	
TREVES.				
Ducat, ...	2	5¾	2 26 7	
TURKEY.				
Sequin fonducli, of Cons'ple, 1773,	2	5¾	1 86 8	
Sequin fonducli, of Cons'ple, 1789,	2	5¾	1 84 8	
Half Misseir, 1818, ...		18¼	52 1	
Sequin Fonducli, ...	2	5	1 83	
Yeermeeblekblek, ...	3	1¾	3 02 8	
TUSCANY.				
Zechino, or Sequin, ...	2	5¾	2 31 8	
Ruspone of kingdom of Etruria,	6	17¼	6 93 8	
VENICE.				
Zechino, or Sequin, shares in pro.	2	6	2 31	
WIRTEMBURG.				
Carolin, ...	6	3½	4 89 8	
Ducat, ...	2	5	2 23 5	
ZURICH.				
Ducat, double and half in pro'n,	2	5¾	2 26 7	

FOREIGN WEIGHTS AND MEASURES.

The weights and measures of GREAT BRITAIN are the same as those of the United States, excepting the variations which are noted in the tables of ' Weights and Measures,' page 27.

The weights and measures of FRANCE being more nicely adjusted than those of any other country, will be here given the more fully on that account. It is, however, to be observed, that these weights and measures are according to a *new system*, not yet in very common use.

The fundamental standard adopted in France for the metrical system of weights and measures, is a quadrant of the meridian; that is to say, the distance from the equator to the north pole. This quadrant is divided into ten millions of equal parts, and one of these equal parts is called the METRE, which is adopted as the unit of length, and from which by decimal multiplication and division all other measures are derived.

In order to express the decimal proportions, the following vocabulary of names has been adopted.

For multipliers,

the word *Deca* prefixed, means   10 times.
     " *Hecto* "   "   100 times.
     " *Chilo* "   "   1000 times.
     " *Myria* "   "   10000 times

For divisors,

the word *Deci* prefixed, expresses the 10th part.
     " *Centi* "   "   100th part.
     " *Milli* "   "   1000th part.

It may assist the memory to observe that the terms for multiplying are Greek, and those for dividing, Latin.

Thus, *Deca-metre* means 10 Metres.

Deci-metre " the 10th part of a Metre.
Hecto-metre " 100 Metres.
Centi-metre " the 100th part of a Metre; &c.

FRENCH LONG MEASURE.

The *Metre*, which is the unit of long measure, is equal to 39.371 English inches.

10 milli-metres	= 1 centi-metre,
10 centi-metres	= 1 deci-metre,
10 deci-metres	= 1 METRE
10 Metres . .	= 1 deca-metre,
10 deca-metres	= 1 hecto-metre,
10 hecto-metres	= 1 chilo-metre,
10 chilo-metres	= 1 myria-metre.

FRENCH SQUARE MEASURE.

The *Are*, which is a square deca-metre (or 100 square Metres), is the unit of square or superficial measure, and is equal to 3.953 English square rods.

10 milliares . .	= 1 centiare;
10 centiares . .	= 1 deciare;
10 deciares . .	= 1 ARE;
10 Ares . .	= 1 decare;
10 decares . .	= 1 hectare;
10 hectares . .	= 1 chilare;
10 chilares . .	= 1 myriare.

FRENCH MEASURES OF CAPACITY.

The *Litre*, which is the cube of a decimetre, is the unit of all liquid measures, and of all other measures of capacity. The Litre is equal to 61.028 English cubic inches.

10 millilitres . .	= 1 centilitre;
10 centilitres . .	= 1 decilitre;
10 decilitres . .	= 1 LITRE;
10 Litres . .	= 1 decalitre;
10 decalitres . .	= 1 hectolitre;
10 hectolitres . .	= 1 chilolitre;
10 chilolitres . .	= 1 myrialitre.

FRENCH SOLID MEASURE.

The *Stere*, which is a cube of the metre, is the unit of solid measure, that is used for fire-wood, stone, &c. The Stere is equal to 35.31714 English cubic feet; it is the same as the chilolitre in measures of capacity.

10 decisteres .   =1 Stere;
10 Steres . .   =1 decastere.

French Weights.

The *Gramme*, which is the weight of a cubic centimetre of distilled water of the temperature of melting ice, is the unit of all weights.   The Gramme is equal to 15.434 grains Troy.

		Grains Troy
A milligramme is 1000th part of a gramme,	=	0.0154
A centigramme is 100th part of a gramme,	=	0.1543
A decigramme is 10th part of a gramme,	=	1.5434
A GRAMME	=	15.4340
A decagramme is 10 grammes,	=	154.3400
A hectogramme is 100 grammes,	=	1543.4000
A chilogramme is 1000 grammes,	=	15434.0000
A myriagramme is 10000 grammes,	=	154340.0000

All the preceding French weights and measures are deduced from some decimal proportion of the *metre*.   Thus the chilogramme corresponds with the contents of a cubic vessel of pure water at the lowest temperature, the side of which vessel is the *tenth* part of the metre (the decimetre), and the gramme answers to the like contents of a cubic vessel, the side of which is the *hundredth* part of the metre (the centimetre); for the contents of all cubic vessels are to each other in the triplicate ratio of their sides.

100 lb. of HAMBURGH	=106.8 lb. avoirdupois.
The shipfund is 280 lb.	=299 lb. avoirdupois.
1 foot, Hamburgh	=11.289 inches, U. S.
The Hamburgh ell is 2 feet	=22.578 inches, U. S.
The Hamburgh mile	=4.684 miles, U. S.
The fass of Hamburgh	=1.494 bushel of U. S.
The last of grain is 60 fasses	=89.64 bushels of U. S.
The ahm of Hamburgh	=38.25 gallons, U. S.

100 lb. of AMSTERDAM	=108.93 lb. avoirdupois.
4 shipfunds is 1 ship-pound	=326.79 lb. avoirdupois.
The Amsterdam last	=85.248 bushels, U. S.

The Aam (liquid)   =41 gallons, U. States.
The Amsterdam foot   =11.147 inches, U. S.
The ell of Amsterdam   =27.0797 inches, U. S.
The ell of the Hague   =27.333 inches, U. S.
The ell of Brabant   =27.585 inches, U. S.

100lb. of PORTUGAL   =101.19 lb. avoirdupois.
An arroba is 32 lb.   =32.38 lb. avoirdupois.
The moyo, a dry measure   =23.03 bushels, U. S.
The almude, a liquid measure   =4.37 gallons, U. S.
The pe or foot, long measure   =12.944 inches, U. S.
The palmo or standard span   =8.64 inches, U. S.
The vara is 5 palmos   =43.2 inches, U. S.
The Portuguese mile   =1.25 mile, U. S.

100 lb. of SPAIN   =101.44 lb. avoirdupois.
The arroba of wine   =4.245 gallons, U. S.
The fanega, $\frac{1}{12}$ of a cahiz =1.599 bushels, U. S.
The Spanish standard foot   =11.128 inches, U. S.
The vara, a cloth measure   =33.384 inches, U. S.
The legua or league   =4.291 miles, U. S.

100lb. victualie, of SWEDEN=93.76 lb. avoirdupois.
The Swedish foot   =11.684 inches, U. S.
The Swedish ell is 2 feet   =23.368 inches, U. S.
The Swedish mile   =6.64 miles, U. S.
The kann, (both dry and liquid)=159$\frac{2}{3}$ cubic in. U. S.
100 kanns   =69.09 galls. wine, U. S.
100 kanns   =7.42 bushels, U. S.

100 lb. of RUSSIA   =90.26 lb. avoirdupois.
400 lb. make 1 berquit   =361.04 lb. avoirdupois.
A pood is 40 lb. Russian   =36.1054 lb. avoir's.
A chetwert, a dry measure,   =5.952 bushels, U. S.
The vedro, a liquid measure,   =3.246 gallons, U. S.
The Russian inch   =1 inch, U. S.
The Russian foot   =13.75 inches, U. S.
The arsheen, a cloth measure,   =28 inches, U. S.
The sashine or fathom   =7 feet, U. S.
A werst or Russian mile   =3500 feet, U. S.

22

100 lb. of **PRUSSIA**	=103.11 lb. avoirdupois.
The quintal is 110 lb.	=113.421 lb. avoir's.
The scheffel, a dry measure,	=1.5594 bushel, U. S.
The eimer, a liquid measure,	=18.14 gallons, U. S.
The Prussian foot	=12.356 inches, U. S.
The Prussian ell	=26.256 inches, U. S.
The Prussian mile	=4.68 miles, U. S.

100 lb. **DENMARK,**	=110.28 lb. avoir's.
The centner is 100 lb.	=110.28 lb. avoir's.
The shippond is 320 lb.	=352.896 lb.
The bbl.or toende, a dry meas.	=3.9472 bushels, U. S.
The viertel, a liquid measure.	=2.041 gallons, U. S.
The Danish or Rhineland foot	=12.356 inches, U. S.
The Danish ell is 2 feet	=24.712 inches, U. S.
The Danish mile	=4.684 miles, U. S.

A cantaro grosso, **NAPLES,**	=196.5 lb. avoirdupois.
The cantaro piccolo	=106 lb. avoirdupois.
The tomolo, a dry measure,	=1.451 bushels, U. S.
The carro is 36 tomoli	=52.236 bushels, U. S.
The barile, a liquid measure,	=11 gallons. U. S.
The carro of wine is 24 barili	=264 gallons. U. S.
The palmo, long measure,	=10.38 inches, U. S.
The canna is 8 palmi	=83.04 inches, U. S.

100 lb. or libras, **SICILY,**	=70 lb. avoirdupois.
The cantaro grosso	=192.5 lb. avoirdupois.
The cantaro sottile	=175 lb. avoirdupois.
The salma grossa, a dry measure,	=9.77 bushels, U. S.
The salma generale	=7.85 bushels, U. S.
The salma, a liquid measure,	=23.06 gallons, U. S.
The palmo, a long measure,	=9.5 inches, U. S.
The canna is 8 palmi	=76 inches, U. S.

100 lb. of **LEGHORN,**	=75 lb. avoirdupois.
The sacco, a dry measure,	=$2\frac{1}{16}$ bushels, U. S.
The barile, a liquid measure,	=12 gallons, U. S.
155 braccia, cloth measure,	=100 yards, U. S.
The canna of 4 braccia	=93 inches, U. S.

100 lb. peso grosso of GENOA, = 76.875 lb. avoir's.
100 lb. peso sottile   = 69.89 lb. avoir's.
The mina, a dry measure,   = 3.426 bushels, U. S.
The mezzarola, liquid measure, = 39.22 gallons. U. S.
The palmo, long measure,   = 9.725 inches, U. S.
The braccio is 2⅓ palmi   = 22.692 inches, U. S.

100 lb. peso grosso, VENICE, = 105.18 lb. avoir's.
100 lb. peso sottile   = 66.4 lb. avoir's.
The stajo, a dry measure,   = 2.27 bushels, U. S.
The moggio is 4 staja   = 9.08 bushels, U. S.
The bigoncia, liquid measure, = 34.2375 galls. U. S.
The anfora is 4 bigonzi.   = 136.95 galls. U. S.
The braccio for woollens,   = 26.61 inches, U. S.
The braccio for silks   = 24.8 inches, U. S.
The Venetian foot   = 13.68 inches, U. S.

100 lb. of TRIESTE,   = 123.6 lb. avoirdupois.
The stajo, dry measure,   = 2.344 bushels, U. S.
The orna, or eimer, liquid   = 14.94 gallons, U. S.
The ell for woollens   = 26.6 inches, U. S.
The ell for silks   = 25.2 inches, U. S.
The Austrian mile   = 4.6 miles, U. S.

100 lb. or libras, ROME,   = 74.77 lb. avoirdupois.
The rubbio, dry measure,   = 8.356 bushels, U. S.
The barile, liquid measure,   = 15.409 galls. U. S.
The Roman foot   = 11.72 inches, U. S.
The mercantile canna   = 78.34 inches, U. S.
The Roman mile   = 7.4 furlongs, U. S.

100 lb. or 100 rottoli, MALTA, = 174.5 lb. avoirdupois.
The salma, dry measure,   = 8.221 bushels, U. S.
The foot of Malta   = 11⅙ inches, U. S.
The canna is 8 palmi   = 81.9 inches, U. S.

The cantaro, kintal, SMYRNA, = 129.48 lb. avoirdupois.
The oke or oka   = 2.833 lb. avoirdupois.
The killow, dry measure,   = 1.456 bushels, U. S.
The pic, long measure,   = 27 inches, U. S.

A factory maund of BENGAL, = 74⅔ lb. avoirdupois.
A bazar maund, = 82²/₁₅ lb. avoirdupois.
The haut or cubit = 18, inches, U. S.
The guz = 1 yard U. S.
The coss or mile = 1.238 miles U. S.

The maund of BOMBAY, = 28 lb. avoirdupois.
The candy is 20 maunds = 560 lb. avoirdupois.
A bag of rice weighs 6 maunds = 168 lb. avoirdupois.
The candy, dry measure, = 25 bushels, U. S.
The haut or covid = 18 inches, U. S.

The maund of MADRAS, = 25 lb. avoirdupois.
The candy is 20 maunds = 500 lb. avoirdupois.
The baruay, a Malabar weight, = 482.25 lb. avoir's.
The garee, dry measure, = 140 bushels, U. S.
The covid, long measure, = 18 inches, U. S.

The pecul of CANTON, = 133⅓ lb. avoirdupois.
The catty is 100th part of a pecul, = 1.333 lb. avoirdupois.
The covid or cobre, long meas. = 14.625 inches U. S.

The pecul of JAPAN, = 130 lb. avoirdupois.
The catti is 100th part of a pecul, = 1.3 lb. avoirdupois.
The inc or tattamy, long meas. = 6.25 feet, U. S.

The bahar of BENCOOLEN, = 560 lb. avoirdupois.
The bamboo, liquid measure, = 1 gallon, U. S.
The coyang is 800 bamboos = 800 gallons, U. S.

The bahar of ACHEEN, = 423.425 lb. avoir's.
The maund of rice = 75 lb. avoirdupois.
The loxa of betel nuts = 10000 nuts.
The loxa of nuts (when good) = 168 lb. avoirdupois.

The pecul of BATAVIA, = 135¹⁰/₁₆ lb. avoirdupois.
33 kannes, liquid measure, = 13 gallons, U. S.
The ell, long measure, = 27 inches, U. S.

The candy of COLOMBO, = 500 lb. avoirdupois.

XXXIX.

MENSURATION.

MENSURATION is the art or practice of measuring, and has primary reference to the measurement of superficies and solids.

Mensuration involves a knowledge of Geometry; and, as that science is not the object of this work, we shall confine our exercises under this head to those measurements, which are most likely to be useful in the ordinary concerns of life.

SUPERFICIES OR SURFACE.

It has already been taught, that surfaces are measured in *squares*, and that the area of any square figure, or any parallelogram is found by multiplying together the length and breadth of the figure. For observations on the square and parallelogram, see page 162.

AREA OF A RHOMBUS. A rhombus is a figure with four equal sides, having two of its angles greater, and two less than the angles of a square. The greater angles are called *obtuse* angles, ,and the smaller, *acute* angles. To find the area of a rhombus, *first drop a perpendicular from one of the obtuse angles to the opposite side, then multiply the side by the perpendicular.*

1. How many square feet are there in a flooring, the form of which is that of a rhombus, measuring 15 feet on the side, and 12.5 feet in the perpendicular?

AREA OF A RHOMBOID. A rhomboid is a figure with four sides, which are not all equal, but whose *opposite* sides are equal, and whose opposite angles are

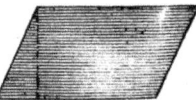

22*

equal, having, like a rhombus, two obtuse, and two acute angles. To find the area of a rhomboid, *drop a perpendicular from one of the obtuse angles, to the opposite longer side, and multiply the longer side by the perpendicular.*

2. What is the area of a rhomboid whose longer side is 18.75 feet, and whose perpendicular is 9.25 feet?

AREA OF TRIANGLES. It is obvious, that a right-angled triangle contains just half as much surface as would be contained in a square or parallelogram, two of whose sides are formed by the base and perpendicular of the triangle. Therefore, *the area of a right-angled triangle is found, by multiplying together either the base and half the perpendicular, or, the perpendicular and half the base.*

3. How many square rods of land are there in a lot, which is laid out in a right-angled triangle, the base measuring 19 rods, and the perpendicular 15 rods?

4. How many acres of land in a lot, whose form is that of a right-angled triangle, the base measuring 113 rods, and the perpendicular 75 rods?

An *Equilateral* triangle is a triangle whose sides are all equal—such is the first of the two triangles adjoined. An *obtuse-angled* triangle is that which has one obtuse angle —such is the second of the triangles adjoined. Whatever may be the form of a triangle, if it have not a right angle, it must be cut into two right-angled triangles before it can be measured: and this is done by dropping a perpendicular from the opposite angle to the

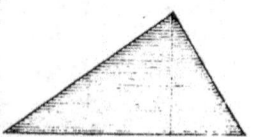

base. *The area is then found by multiplying together the base and half the perpendicular, or, the perpendicular and half the base.*

5. How many square inches in a triangle, whose base is 17¼ inches, and whose perpendicular height is 11¾ inches ?

6. How many square feet in a board 18 feet long, 16 inches wide at one end, and tapering to a point at the other end ?

7. How many square feet in a plank 14 feet long, 17 inches wide at one end, and 10 inches wide at the other end ?

In this example, add the width of the two ends together, and take half the sum for one of the factors.

AREA OF CIRCLES. To find the area of a circle, *multiply the circumference by half the diameter, and divide the product by 2.* When either the circumference, or the diameter is the only dimension known, the other dimension may be found, as stated in page 173.

8. What is the area of the head of a cask, the diameter of which is 18 inches ?

9. Suppose a cylinder to measure 3 feet in circumference; what is the area of one end ?

AREA OF GLOBES. To find the convex area of a globe or sphere, *multiply the circumference and diameter together.* When the diameter is not known, it may be found from the circumference, as stated in page 173.

10. How many square inches are there on the surface of a globe, whose circumference is 14 inches ?

11. Suppose the earth to be 25020 miles in circumference, what must be the area of its whole surface ?

SOLIDS AND CAPACITIES.

It has already been taught, that solids and capacities are measured in cubes. It has also been shown, that the contents of any thing having six sides— its opposite sides being equal, and all its angles being right angles— are found by multiplying together the length, and breadth, and depth of the thing.

SOLIDITY OF WEDGES. To find the solid contents of a wedge, *first, find the area of the head or end of the wedge, and then multiply this area by half the length.*

12. How many solid inches are there in a wedge, 12 inches long, 3 inches wide, and 1½ inch thick at the head?

13. What are the solid contents (in feet and inches) of a plank, 15 feet long, 17 inches wide, 2¼ inches thick at one end, and the thickness tapering to nothing at the other end?

14. What are the solid contents of a stick of hewn timber, measuring in length 13 feet, in breadth 2 ft. 4 in., in depth 2 feet at one end, and 1 ft. 6 in. at the other end?

In this example, add the depth of the two ends together, and take one half of the sum for the depth to be used in the multiplication.

SOLIDITY OF PRISMS. A prism is a body with two equal ends, which are either square, triangular, or polygonal, and three or more sides, which meet in parallel lines, running from the several angles of one end to those of the other. The adjoined is a representation of a triangular prism.

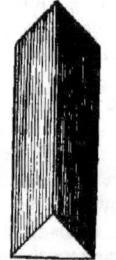

The solid contents of prisms of all kinds, whether square, triangular, or polygonal, are found by one general rule, viz. *Find the area of the end or base, and multiply this area by the length or height.*

15. How many cubic inches are there in a triangular prism, which is 16 inches in length, the ends measuring 1.2 inches on a side, and 1.01 inches perpendicular?

16. How many cubic feet are there in a stick of timber 18 feet long, hewn 3 square, the ends forming equilateral triangles of 10 inches side, and 8.7 inches perpendicular?

SOLIDITY OF CYLINDERS. A cylinder is a round body, the two opposite sides, or ends of which, are circular planes, equal, and parallel. For instance, a stick of round timber of uniform circumference, having its ends sawed at right angles with its length, is a cylinder: also, a common grindstone is a cylinder. To find the solid contents of a cylinder, *first, find the area of one end, and then multiply this area by the length.*

17. What are the solid contents of a cylinder whose length is 5 feet, and circumference 6.4 feet? (To find the diameter, see page 173.)

18. What are the contents of a cylinder whose length is 2 feet, and diameter 10 inches?

SOLIDITY OF PYRAMIDS. Solids, which decrease gradually from the base, till they come to a point, are called *pyramids.* They are of different kinds, according to the figure of their bases. If the pyramid has a square base, it is called a *square pyramid;* if a triangular base, a *triangular pyramid;* if the base be a circle, a *circular pyramid,* or a CONE. The point in which the pyramid ends is called the *vertex.* A line through the centre of the pyramid, from the vertex to the base, is the *height.*

The *Frustrum* of a pyramid is what remains, after any portion of the top has been cut òff, parallel to the base.

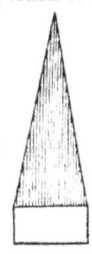 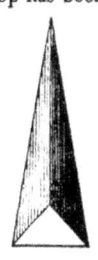 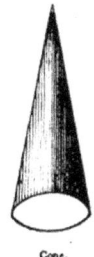 

Square Pyramid.   Triangular Pyramid   Cone.   Frustrum.

To find the cubical contents of a pyramid, *first find the area of the base, then multiply this area by one-third of the height.*

19. How many cubic inches are there in a square pyramid, 3 feet in height, and 9 inches square at the base?

20. How many cubic inches are there in a triangular pyramid, measuring 4 feet in height, 12 inches on each side of the base, and 10.4 inches from either angle of the base perpendicular to the opposite side?

21. How many cubical inches in a cone, the height of which is 19 inches, and the diameter of the base 12 inches?

SOLIDITY OF FRUSTRUMS. To find the cubical contents of the frustrum of a square pyramid, *multiply the side of the base by the side of the top, and to the product add one-third of the square of the difference of the sides, and the sum will be the mean area between the two ends. Multiply the mean area by the height, and the product will be the cubical contents.*

To find the cubical contents of the frustrum of a Cone, *multiply together the diameters of base and top, and to the product add one-third of the square of the difference of the diameters; then multiply this sum by .7854, and the product will be the mean area between the two ends. Multiply the mean area by the height, and the product will be the cubical contents.*

22. How many cubical inches in the frustrum of a square pyramid, 20 inches in height, 12 inches square at the base, and 5 inches square at the top?

23. How many cubic feet in a stick of hewn timber, 18 feet long, 16 inches square at one end, and 12½ inches square at the other end?

24. How many cubic inches are there in the frustrum of a cone, measuring 3 feet in height, 16 inches in diameter at the base, and 6 inches in diameter at the top?

25. How many gallons of water can be contained in a round cistern, 6 feet in height, 4 feet in diameter at the bottom; and 3½ feet in diameter at the top? (Allow 231 cubic inches to the gallon.)

SOLIDITY OF GLOBES. To find the cubical contents of a globe or sphere, *first, find the convex area, as before directed, then multiply the area by one-sixth of the diameter; the product will be the cubical contents.*

26. What are the cubical contents of a globe measuring 25 inches in circumference?

27. How many cubic miles does the earth contain, allowing its circumference to be 25020 miles?

SOLIDITY OF IRREGULAR BODIES. The cubical contents of a body, which cannot be reduced to regular geometrical form may be found as follows. *Immerse it in a vessel partly full of water; then the contents of that part of the vessel filled by the rising of the water will be the contents of the body immersed.*

28. How many cubic inches are there in a lobster, which, being immersed in a bucket 10 inches in diameter at top and bottom, raises the water 3 inches?

GAUGING OF CASKS.

Although the difficulty of getting the true dimensions of the interior of casks, and the variety of their curve, must prevent perfect accuracy in their mensuration, yet, by careful observation in taking the dimensions, a result may be had, which will be sufficiently correct for all common purposes.

RULE. *Take the interior length of the cask, the diameter at the bung, and the diameter at the head, all in inches. Subtract the head diameter from the bung diameter, and note the difference.*

If the staves of the cask be MUCH *curved between the bung and head, multiply the difference noted by .7; if but* LITTLE *curved, by .6; or, if they be of a* MEDIUM *curve, by .65; and add the product to the head diameter; the sum is the mean diameter, and thus the cask is reduced to a cylinder.*

Square the mean diameter, and multiply the square by the length of the cask; then divide this product by 294, and the quotient will be the number of wine gallons, which the cask may contain.

It may be observed, that when a cask is *reduced to a cylinder*, its contents may be found in cubical inches, and thence its contents in bushels, or any other of the measures of capacity.

The length of the cask is most conveniently taken by callipers; allowing for the thickness of both heads, from 1 to 2 inches, according to the size of the cask. When no callipers can be had, the length of the stave must be taken *in a right line*, and a proper deduction made for the chimes, with that for the heads. The head diameter is to be taken within the chimes, and from .3 to .6 of an inch must be deducted, on account of the greater thickness of the stave inside the head.

29. How many gallons will a cask contain, the interior of which measures 34.5 inches in length, 19 inches in diameter at the bung, and 16 inches in diameter at the head; the staves being much curved?

30. How many gallons will a cask contain, the dimensions of which are 43 inches in length, 31.4 inches bung diameter, and 26 inches head diameter; the staves being but little curved?

31. Find the capacity of a cask measuring 52 inches in length, 33.5 inches bung diameter, 25.3 inches head diameter, and of medium curve between the bung and head.

TONNAGE OF VESSELS.

There are two methods of measuring a vessel practised —one by the ship-carpenter, who builds the vessel at a certain price per ton, and another by the officers of government, who collect the revenue.

CARPENTERS' RULE. *For single-decked vessels, multiply together the length of the keel, the breadth at the main beam, and the depth of the hold—all in feet—and divide the product by 95; the product is the tonnage. For double-decked vessels, take half the breadth at the beam for the depth of the hold, and work as before.*

When a single-decked vessel has its deck bolted at any height above the wale, the carpenter is usually paid for one-half of this extra height; that is, one-half of the height above the wale is added to the depth below the wale, and this sum is used in the calculation, as the depth of the hold.

GOVERNMENT RULE. *"If the vessel be double-decked, take the length thereof from the fore part of the main stern, to the after part of the stern-post, above the upper deck; the breadth thereof at the broadest part above the main wales, half of which breadth shall be accounted the depth of such vessel, and then deduct from the length, three-fifths of the breadth, multiply the remainder by the breadth and the product by the depth, and divide this last product by 95, the quotient whereof shall be deemed the true contents or tonnage of such ship or vessel; and if such ship or vessel be single-decked, take the length and breadth, as above directed, deduct from said length three-fifths of the breadth, and take the depth from the under side of the deck plank to the ceiling in the hold, then multiply and divide as aforesaid, and the quotient shall be deemed the tonnage."*

32. What is the carpenter's tonnage of a single-decked vessel, the keel of which measures 60 feet, the breadth 20 feet, and the depth 8 feet ?

33. What is the carpenter's tonnage of a double-decked vessel of 72 feet keel, and 22.5 feet breadth ?

34. A merchant agreed with a carpenter to build a single-decked vessel of 58 feet keel, 20 feet breadth at the beam, and 8 feet hold, but afterwards chose to make the hold 10 feet deep, by raising the deck 2 feet above the wale. What tonnage must be paid for ?

35. What is the government tonnage of a double-decked vessel, 110.5 feet keel, and 30.6 feet breadth at the beam ?

36. What is the government tonnage of a single-decked vessel, which measures 76.4 feet in length, 28.6 feet in breadth, and 12.3 feet in depth ?

37. What is the government tonnage of a single-decked vessel, whose length is 66 feet, breadth 20 feet, and depth 9 feet ?

XL.

MECHANICAL POWERS.

The MECHANICAL POWERS are certain simple instruments employed in raising greater weights, or overcoming greater resistance than could be effected by the direct application of natural strength. They are usually accounted six in number; viz. the *Lever*, the *Wheel and Axle*, the *Pulley*, the *Inclined Plane*, the *Wedge*, and the *Screw*.

The advantage gained by the use of the mechanical powers, does not consist in any increase of the quantum of force exerted by the moving agent, but, in the *concentration* of force; that is, in bringing the whole force of a power acting through a greater space, into an action within a less space. The principle is illustrated by the consideration, that the quantum of force necessary to raise 1 pound 10 feet, will raise 10 pounds 1 foot.

Weight and *Power*, when opposed to each other, signify *the body to be moved* and *the body that moves it*.

THE LEVER.

A lever is any inflexible bar, which serves to raise weights, while it is supported at a point, which is the centre of its motion, by a  *fulcrum* or prop. There are several kinds of lever used in mechanics; the more common kind, however, is that which is shown above.

As the distance between the weight and fulcrum is to the distance between the power and fulcrum, so is the power to the weight.

It must be observed, that, in the above proportion, and in all the succeeding proportions of weight and power, the power intended is only sufficient to *balance* the weight. If the weight is to be *raised*, sufficient power must be added to overcome friction; then any further addition of power will produce motion; and the comparative velocity of the weight and power, will depend on the comparative length of the two arms of the lever. It is a universal principle in mechanics, that the ratio of the *power* to the *weight* is equal to the ratio of the *velocity* of the weight to the *velocity* of the power.

1. If a man weighing 160 pounds rest on the end of a lever 10 feet long, what weight will he balance on the other end, the fulcrum being 1 foot from the weight?

In this example, the distance between the weight and fulcrum being 1 foot, that between the power and fulcrum is 10—1=9 ft. Then 1 ft. : 9 ft.=160 lb. : A

2. Suppose a weight of 1440 pounds is to be raised with a lever 10 feet long, the fulcrum being fixed 1 foot from the weight; what power must be applied to the other end of the lever, to effect a balance?

(9 ft. : 1 ft.=1440 lb. : A)

3. If a weight of 1440 pounds be placed 1 foot from the fulcrum; at what distance from the fulcrum must a power of 160 pounds be placed, to balance the weight?

(160 lb. : 1440 lb.=1 ft. : A)

4. At what distance from a weight of 1440 pounds must the fulcrum be placed, so that a power of 160 pounds, applied 9 feet from the fulcrum, will effect a balance?

(1440 lb. : 160 lb.=9 ft. : A)

5. If one arm of a lever be 44 feet, and the other 5 feet, what power must be applied to the longer arm, to balance a weight of 500 pounds on the shorter arm?

6. Suppose a lever 6 feet long, with one end applied to a rock, which weighs 1000 pounds, and resting on a fulcrum 1½ foot from the rock; what power must be applied to the other end, to balance the rock?

7. Suppose a bar 12 feet long to have 60 pounds attached to one end, and 30 pounds to the other, at what distance from each end must a fulcrum be placed, to produce a balance?

8. If A and B carry a weight of 250 pounds, suspended upon a pole between them, 5 feet from A, and 3 feet from B, how many pounds does each carry?

THE WHEEL AND AXLE.

Tne wheel and axle are
here represented, with the
weight attached to the circumference of the axle, and
the power applied to the circumference of the wheel. The
principle of the lever is obvious in the wheel and axle—
the axis or common centre
being the fulcrum, the circumference of the wheel being the
power end of the lever, and
the circumference of the axle, [
the end applied to the weight.　Hence, the radius of the
axle is to the radius of the wheel, as the power is to the
weight: or, by a statement more frequently convenient—

*As the diameter of the axle is to the diameter of the
wheel, so is the power to the weight.*

9. A mechanic would make a windlass in such manner,
that 1 pound applied to the wheel, shall he equal to 10
pounds suspended from the axle. Now, supposing the
axle to be six inches in diameter, what must be the diameter of the wheel?

10. Suppose the diameter of a wheel to be 8 feet, what
must be the diameter of the axle, that 1 pound on the
wheel shall balance 15 pounds on the axle?

11. Suppose the diameter of an axle to be 4 inches,
and that of the wheel 3 feet; what power at the wheel
will balance 28 pounds at the axle?

12. If the diameter of a wheel be 7 feet, and that of
the axle 8 inches, what weight at the axle will balance
40 pounds at the wheel?

13. There are two wheels; one of which is 6 feet in
diameter, with an axle of 9 inches diameter; and the other
is 4 feet in diameter, with an axle of 7 inches diameter.
Suppose the power cord of the smaller wheel to be coiled
upon the axle of the larger; what weight on the axle of
the smaller wheel would be balanced 100 at the power
cord of the larger wheel?

THE PULLEY.

A pulley is a small wheel, which turns on an axis passing through its centre and fixed in a block, receiving its motion from a cord, that passes round its circumference. The pulley is either single or combined: it is also, either fixed, or movable. If a power sustain a weight by means of a single, fixed pulley—a cord passing over it, with the weight attached to one end and the power to the other—the power and weight are equal: and if the pulley be put in motion, the velocity of the power, and the velocity of the weight will also be equal. But, if the fixed pulley be combined with one movable pulley—as represented in the first set of pulleys above—the weight is equal to twice the power which sustains it; and if the pulley be put in motion, the velocity of the power will be equal to twice the velocity of the weight. Thus, every cord going over a movable pulley, adds 2 to the powers, and hence, in a system of pulleys, we have the following proportion.

As 1 *is to twice the number of movable pulleys, so is the power to the weight.*

14. In the second set of pulleys represented above, three of the pulleys are fixed, and three are movable. If a power of 45 pounds were applied to the cord, what weight would it balance?

15. What power must be applied to a cord that runs over 2 movable pulleys, in order to balance a weight of 800 pounds?

16. What power must be applied to a cord that runs over 6 movable pulleys, to balance a weight of 2000 pounds?   23*

17. If a cord, which runs over 3 movable pulleys, be attached to an axle 4 inches in diameter, the wheel of the axle being 38 inches in diameter, and a power of 20 pounds be exerted at the circumference of the wheel, what weight would be raised under the pulleys?

THE INCLINED PLANE.

An inclined plane is a plane making an angle with the horizon. For instance, a plank presents an inclined plane, when one end is resting upon the level ground, and the other end is raised to any height less 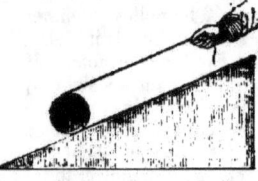 than that which would render it vertical. A convenient use of the inclined plane is exemplified in rolling casks from a cellar, upon sloping pieces of timber, or planks.

On an inclined plane, as the perpendicular height of the plane is to the length of the plane, so is the power to the weight.

18. A certain inclined plane is 16 feet in length, and 7 feet in perpendicular height. What weight might be drawn up this plane, by a power, which, if exerted on a cord over a single, fixed pulley, would raise 25 pounds?

19. What power would be necessary to sustain a rolling weight of 1000 pounds, upon an inclined plane of 75 feet length, and 38 feet perpendicular height?

20. What must be the length of an inclied plane, whose perpendicular height is 15 feet, that the exertion of the power of 42 pounds shall draw up 200 pounds?

21. On a rail-road, there is an inclined plane of 80 rods in length, rising to a perpendicular height of 50 feet. What power must be exerted on the summit, to draw up a train of cars weighing 62000 pounds?

22. Suppose a set of pulleys, 3 of which are movable, to be applied to a weight upon an inclined plane of 50 feet length, and 14 feet perpendicular height; what weight upon the plane, would be sustained by 40 pounds at the power cord of the pulleys?

THE WEDGE.

The wedge may be viewed as a moving inclined plane; the head of the wedge, where the power is applied, answering to the perpendicular height of the plane. In the wedge, however, the inclined plane is double, and the force produced by its advance is divided into two equal parts, acting at right angles with each side.

As the breadth of the head of a wedge is to the length of its side, so is the power acting against the head, to the force produced at the side.

Observe, that the force mentioned in the above proportion, respects one side of the wedge, only. If the forces against both sides be required, then, only half the breadth of the head must be taken into the proportion.

In the common mode of applying the wedge, the friction against the sides is very great— at least equal to the force to be overcome. Therefore, not less than one-half of the power is lost; and for this loss there is no allowance made in the above proportion. The wedge, however, has a great advantage over all the other mechanical powers, arising from the force of percussion or blow with which the head is struck, by a mallet. The power thus obtained is incomparably greater than that of any dead weight or pressure, such as is commonly employed on other instruments.

23. Suppose a power of 50 pounds to be applied to a wedge, the head of which is 2 inches broad, and the side 12 inches long, what weight of force would be effected on either side; if there were no friction to resist?

24. If a force of 1000 pounds is to be effected on the side of a wedge, that is 14 inches long, and 3 inches broad at the head, what power must be applied to the head; allowing nothing for friction? Again, allowing the friction, which is to be overcome, to be equal to the force effected, what power will be necessary?

THE SCREW.

The screw is a spiral thread or groove, cut round a cylinder, and every where making the same angle with the length of the cylinder. In one round of the spiral, it rises along the cylinder, the distance between two threads. Therefore, if the surface of the cylinder, with the spiral thread on it, were unfolded and stretched into a plane, the spiral would form a 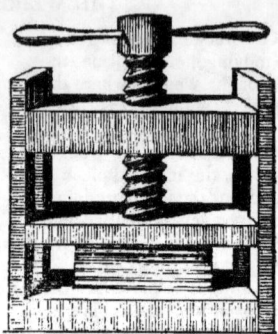 straight inclined plane, whose length would be to its height, as the circumference of the cylinder is to the distance between two threads of the screw. The inclined plane being thus recognised in the screw, the following proportion is obvious.

As the distance between two threads of a screw is to the circumference of the circle described by one revolution of the power, so is the power to the weight.

The length of the lever to which the power is applied, being one-half of the diameter of the circle round which the power revolves, the circumference may be found from the lever, as taught in page 173.

In the common use of the screw, about one-third of the power is expended in overcoming friction; and for this loss, no allowance is made in the above stated proportion.

25. If the threads of a screw be 1 inch apart, and a power of 50 pounds be exerted at the end of a lever 70 inches long, what weight of force will be produced at the end of the screw; allowing nothing for friction.

26. If the threads of a screw be .2 of an inch apart, and a power of 40 pounds be exerted at the end of a lever 30 inches long, what will be the force at the end of the screw; allowing ⅓ of the power to be lost in overcoming friction?

27. Suppose a power of 48 pounds is to be employed to effect the weight of 5000 pounds, by means of a screw, whose threads are 1.3 inches apart; what must be the length of the lever; allowing $\frac{1}{3}$ of the power to be lost in overcoming friction?

28. Suppose the end of a screw, whose threads are .8 of an inch apart, and whose lever is 7ft. long, to be set upon a wedge, that is 15in. long at the side, and 2 inches broad at the head; what weight of force would be effected on either side of the wedge, by applying 100 pounds' power to the lever; allowing $\frac{1}{3}$ of the force on the screw, and $\frac{1}{2}$ of that on the wedge to be lost in friction?

XLI.

MISCELLANEOUS QUESTIONS.

1. What vulgar fraction is that, which being multiplied by 15, will produce $\frac{3}{4}$?

2. What decimal fraction is that, which being multiplied by 15, will produce .75?

3. What quantity is that, which being divided by $\frac{3}{4}$, gives the quotient 21?

4. What vulgar fraction is that, from which if you take $\frac{3}{5}$, the remainder will be $\frac{1}{8}$?

5. What vulgar fraction is that, to which if you add $\frac{2}{5}$, the sum will be $\frac{5}{6}$?

6. What quantity is that, which being multiplied by $\frac{2}{3}$, produces the fraction $\frac{1}{4}$?

7. What quantity is that, from which if you take $\frac{2}{5}$ of itself, the remainder will be 12?

8. What quantity is that, to which if you add $\frac{2}{5}$ of $\frac{3}{11}$ of itself, the sum will be 61?

9. A farmer carried to market a load of produce, consisting of 780lb. of pork, 250lb. of cheese, and 154lb. of butter; he sold the pork at 6 cents, the cheese at 8 cents, and the butter at 15 cents per lb.; and agreed to take in pay, 60lb. of sugar at 10 cents per lb., 15 gallons

of molasses at 40 cents a gallon, ½ barrel of mackerel at $3.50, 4 bushels of salt at 90 cents a bushel, and the balance in cash. How much money did he receive?

10. A and B commenced business with equal sums of money; A gained a sum equal to ⅓ of his stock, but B lost $200, and then had only half as much as A. What was the original stock of each?

11. A man was hired for a term of 50 days on conditions, that for every day he worked he should receive 75 cents, and for every day he was idle he should pay 25 cents for his board; at the expiration of the time, he was entitled to $27.50. How many days was he idle?

12. A and B have the same income; A saves ⅛ of his; but B, by spending $30. a year more than A, at the end of 8 years finds himself $40. in debt. What is their income, and what does each spend a year?

13. A grocer has two sorts of tea; one at 75 cents a pound, and the other at $1.10 a pound. In what proportion must he mix them, in order to afford the mixture at $1. a pound?

14. A and B can do a piece of work in 5 days; A alone can do it in 7 days. In what time can B do it?

15. After A has travelled 51 miles, B sets out to overtake him, and travels 19 miles to A's 16. How many miles will each have travelled, before B overtakes A?

16. A trader bought a cask of wine, but, in conveying it home, ⅓ of it leaked out. He sold the remainder, at $2.50 a gallon, and thus received what he paid for the whole. How much per gallon did he give for it?

17. A person having spent in one year all his income and ¼ as much more, found that by saving 1/15 of his income afterward, he could, in 4 years make good the deficiency, and have $20 left. What was his income?

18. A young hare starts 40 yards before a grey-hound, and is not perceived by him till she has been up 40 seconds; she scuds away at the rate of 10 miles an hour, and the hound, in view, makes after her at the rate of 18 miles an hour. How long will the course continue, and what will be the length of it, from the place where the hound set out?

19. A man driving his geese to market, was met by another, who said— 'Good morning, with your hundred geese.' He replied— 'I have not a hundred; but if I had half as many more than I have, and two geese and a half, I should have a hundred.' How many had he?

20. If 8 men can build a wall 15 rods long in 10 days, how many men will it take to build a wall 45 rods long in 5 days?

21. A gentleman had £7 17s. 6d. to pay among his laborers; to every boy he paid 6 pence, to every woman 8 pence, and to every man 16 pence; there was one boy to three women, and one woman to two men. What was the number of each?

22. A farmer bought a yoke of oxen, a cow, and a sheep for $82.50; he gave for the cow 8 times as much as for the sheep, and for the oxen 3 times as much as for the cow. How much did he give for each?

23. The head of a fish was 9 inches long, its tail was as long as its head and half its body, and its body was as long as its head and tail both. What was the whole length of the fish?

24. The remainder of a division is 325, the quotient 467, and the divisor is 43 more than the sum of both; what is the dividend?

25. A trader bought a hogshead containing 120 gallons of molasses for 42 dollars. At what price per gallon must he sell it, to gain 15 per cent.?

26. Sold goods to the amount of $3120, to be paid one half in 3 months, and the other half in 6 months. How much must be discounted for present payment, when money is worth 6 per cent. a year?

27. A merchant imported 10 tons of iron at 95 dollars per ton; the freight and duties amounted to 145 dollars, and other charges to 25 dollars. At what price per lb. must he sell, to gain 20 per cent.?

28. The hour and minute hands of a watch are together at 12 o'clock; when are they next together?

29. Suppose two steamboats to start at the same time from places 300 miles apart on the same river; the one proceeding up stream is retarded by the current 2 miles

per hour; the other moving down stream is accelerated the same. If each is propelled by a steam engine, that would move it 8 miles an hour in still water, how far from each starting place will the boats meet?

30. Thomas sold 150 pineapples at $33\frac{1}{3}$ cents apiece, and took no more money than Harry did for watermelons at 25 cents apiece. How much money did each take, and how many melons had Harry?

31. Seven-eighths of a certain number exceeds four-fifths of the same number by 6. What is the number?

32. If 18 grains of silver will make a thimble, and 12 dwt. a teaspoon, how many thimbles and teaspoons, of each an equal number, can be made from 15 oz. 6 dwt.?

33. What are the superficial contents of a piece of wainscot 8 ft. $6\frac{1}{2}$ in. long, and 2 ft. $9\frac{3}{4}$ in. broad?

34. A guardian paid his ward $3500. for $2500. which he had in his hands 8 years. What rate of interest did he allow?

35. A set out from Boston for Hartford precisely at the time, when B at Hartford set out for Boston, distant 100 miles: after 7 hours they met on the road, and it then appeared, that A had ridden $1\frac{1}{2}$ mile an hour more than B. At what rate an hour did each travel?

36. A father divided his fortune among his sons, giving A $4 as often as B 3, and C 5 as often as B 6. What was the fortune, supposing A's share to be $5000.?

37. A prize of 945 dollars is to be divided among a captain, 4 men, and a boy; the captain is to have a share and a half; the men each a share; and the boy $\frac{1}{3}$ of a share. What ought each person to have?

38. A person left 40 shillings to four poor widows; viz. to A he left $\frac{1}{3}$, to B $\frac{1}{4}$, to C $\frac{1}{4}$, and to D $\frac{1}{6}$, desiring the whole might be distributed accordingly. What is the proper share of each?

39. A person looking on his watch, was asked what was the time of day; he answered—It is between 4 and 5, and the hour and minute hands are exactly together. What was the time?

40. Divide 1200 acres of land among A, B, and C, so that B may have 100 acres more than A, and C 64 acres more than B.

41. What length of a board, which is $8\frac{3}{8}$ inches wide, will contain as much as a square foot?

42. What number is that, from which if you take $\frac{3}{7}$ of $\frac{3}{8}$, and to the remainder add $\frac{7}{16}$ of $\frac{1}{26}$, the sum is 10?

43. A can do a piece of work alone in 10 days, and B in 13 days. If both set about it together, in what time will it be finished?

44. A, B, and C were to share $100000. in the proportion of $\frac{1}{3}$, $\frac{1}{4}$, and $\frac{1}{5}$, respectively; but C's part being lost by his death, it is required to divide the whole sum properly between the other two.

45. In an orchard of fruit trees, $\frac{1}{2}$ of them bear apples, $\frac{1}{4}$ pears, $\frac{1}{6}$ plums, and 50 of them cherries. How many trees are there in the orchard?

46. A cistern, containing 60 gallons of water, has 3 unequal faucets for discharging it; the greatest faucet will empty it in one hour, the second in two hours, and the third in three hours. In what time will it be emptied, if they all run together?

47. What sum of money will amount to 336 dollars 42 cents in a year and 4 months, at 6 per cent. per annum, simple interest?

48. A man, when he married, was 3 times as old as his wife; 15 years afterward he was but twice as old as his wife. At what age was each married?

49. Divide 1000 dollars among A, B, and C, so as to give A 120 dollars more, and B 95 dollars less, than C.

50. What fraction is that, to which if $\frac{4}{7}$ of $\frac{3}{5}$ be added, the sum will be 1?

51. A certain cubical stone contains 389017 solid feet. What are the superficial contents of one side?

52. A father dying left his son a fortune, $\frac{1}{4}$ of which he spent in 8 months; $\frac{3}{7}$ of the remainder lasted him 12 months longer; after which he had only 1200 dollars left. How much did his father leave to him?

53. Three travellers met at an inn, and two of them brought their provisions along with them; but the third not having provided any, proposed to the other two, that they should all eat together, and he would pay them for his proportion. This being agreed to, A produced 5 loaves,

and B 3 loaves, which the travellers ate together, and C
paid 8 equal pieces of money as the value of his share,
with which the other two were satisfied, but quarreled
about the division of them. Upon this, the affair was
referred to an umpire, who decided the dispute justly.
What was his decision?

54. What number is that, which being added to $\frac{1}{17}$ of
765, the sum will be equal to the square root of 2601?

55. Two persons talking of their ages, one says, $\frac{2}{3}$ of
my age is equal to $\frac{3}{4}$ of yours, and the difference of our
ages is 10 years. What were their ages?

56. A man bought some lemons at 2 cents each, and
$\frac{3}{4}$ as many at 3 cents each, and then sold them all at the
rate of 5 cents for 2, and thus gained 25 cents. How
many lemons did he buy?

57. There are two cisterns, which are constantly receiving an equal quantity of water; but the first constantly
loses $\frac{1}{6}$ of what it receives. After running 7 days, 10 barrels were taken from the second, and then the quantity of
water in the two was equal. How much water did each
receive per day?

58. A person being asked the hour of the day, said,
the time past noon is equal to $\frac{4}{5}$ of the time to midnight.
What o'clock was it?

59. What number, added to $\frac{1}{61}$ of 3813, will make the
sum 200?

60. A general forming his army into a square, finds he
has 284 soldiers over and above a square; but increasing
each side with one soldier, he wants 25 to fill up a square.
How many soldiers had he?

61. A reservoir for water has two pipes to supply it;
by the first alone it may be filled in 40 minutes, by the
second alone in 50 minutes; and it has a discharging pipe,
by which it may, when full, be emptied in 25 minutes.
Now, if these three pipes were all left open, the influx
and efflux of the water being always at the aforesaid rates,
in what time would the cistern be filled?

62. Three persons do a piece of work; the first and
second together do $\frac{7}{9}$ of it, and the second and third
together do $\frac{7}{11}$ of it. What part of it is done by the
second?

63. A man driving some oxen, some cows, and some sheep, being asked how many he had of each sort, answered, that he had twice as many sheep as cows, and three times as many cows as oxen; and that the whole number was 80. What was the number of each sort?

64. A man has a note of $647. due in 2 years and 7 months without interest; but being in want of money, he will sell the note; what ought he to receive, when interest is 6 per cent. a year?

65. A gentleman bequeathed an estate of $12500. to his wife and son. The son's share was $\frac{7}{9}$ of the wife's share. What was the share of each?

66. A man and his wife found that when they were together, a bushel of corn would last them 15 days; but when the man was absent, it would last the woman alone 27 days. How long would it last the man alone?

67. A farmer sold some calves and some sheep for $108.; the calves at $5. and the sheep at $8. apiece. There were twice as many calves as sheep. What was the number of each sort?

68. A owes B $158.33 due in 11 months and 17 days, without interest, which he proposes to pay at present. What ought he to pay, money being 5 per cent.?

69. At what time, between twelve and one o'clock, do the hour and minute hands of a clock or watch point in directions exactly opposite?

70. If 3 men can do a piece of work in 56 days, and 4 women can do the same in the same time, in what time will one man and one woman together perform it?

71. A son having asked his father's age, the father thus replied; 'your age is 12 years, to which if five-eighths of both our ages be added, the sum will express my age.' What was the father's age?

72. Three gentlemen agree to contribute $730. towards the building of a church at the distance of 2 miles from the first, $2\frac{7}{8}$ miles from the second, and $3\frac{1}{2}$ miles from the third; and they agree, that their shares shall be reciprocally proportional to their distances from the church. How much must each contribute?

73. If A can reap a field in 13 days, and B in 16 days, in what time can both together reap it?

74. A and B set out together from the same place, and travelled in the same direction. A travelled uniformly 18 miles a day, but after 9 days turned and went back as far as B had travelled during those 9 days; he then turned again, and, pursuing his journey, overtook B in $22\frac{1}{2}$ days from the time they first set out. At what rate per day did B uniformly travel?

75. Two men, A and B, are on a straight road, on the opposite sides of a gate; A is distant from it 308 yards, and B 277 yards, travelling each towards the gate. How long must they walk, to make their distances from the gate equal; allowing A to walk $2\frac{1}{3}$ yards, and B 2 yards, per second?

76. I want just an acre of land cut off from the end of a piece, which is $13\frac{1}{2}$ rods wide; how much of the length of the piece will it take?

77. A farmer had oats at 38 cents a bushel, which he mixed with corn at 75 cents a bushel, so that the mixture might be 50 cents a bushel. What were the proportions of the mixture?

78. A grocer mixed 123 lb. of sugar worth 8 cents per lb. with 87 lb. worth 11 cents per lb. and 15 lb. worth 13 cents per lb. What was the mixture worth per lb.?

79. A man travelling from Boston to Philadelphia, a distance of 335 miles, at the expiration of 7 days found that the distance which he had to travel was equal to $\frac{25}{42}$ of the distance, which he had already travelled. How many miles per day did he travel?

80. A gentleman bequeathed an estate of $50000. to his wife, son, and daughter; to his wife he gave $1500. more than to the son, and to his son $3500. more than to his daughter. How much was the share of each?

81. The stock of a cotton manufactory is divided into 32 shares, and owned equally by 8 persons, A, B, C, &c. A sells 3 of his shares to a ninth person, who thus becomes a member of the company, and B sells 2 of his shares to the company, who pay for them from the common stock. After this, what proportion of the whole stock does A own?

82. How many feet in a stock of 18 boards, 12 feet 3 inches long, and 1 foot 8 inches wide?

83. A merchant laid out $50. for linen and cotton cloth, buying 3 yards of linen for a dollar, and 5 yards of cotton for a dollar. He afterwards sold $\frac{1}{4}$ of his linen and $\frac{1}{5}$ of his cotton for $12, which was 60 cents more than it cost him. How much of each did he buy?

84. If 157 dollars 50 cents in 16 months gain 12 dollars 60 cents, in what time will 293 dollars 75 cents gain 11 dollars 75 cents, at the same rate of interest?

85. A merchant having goat-skins, and wishing to get some of them dressed, delivered for that purpose 560 to a currier, to be dressed at $12\frac{1}{2}$ cents each, who agreed to take his pay in dressed skins at 50 cents each. How many dressed skins should the currier return?

86. If eggs be bought at the rate of 5 for 4 cents, how must they be sold per dozen, to gain 25 per cent.?

87. What is the circumference of a wheel, the diameter of which is 5 feet?

88. A lion of bronze, placed upon the basin of a fountain, can spout water into the basin through his throat, his eyes, and his right foot. If he spouts through his throat only, he will fill the basin in 6 hours; if through his right eye only, he will fill it in 2 days; if through his left eye only, in 3 days; if through his foot only, he will fill it in 4 hours. In what time will the basin be filled, if the water flow through all the apertures at once?

89. A man having 100 dollars spent part of it, and afterward received five times as much as he had spent, and then his money was double what it was at first. How much did he spend?

90. A hare starts 50 leaps before a grey-hound, and takes 4 leaps to the hound's 3; but 2 of the hound's leaps are equal to 3 of the hare's. How many leaps must the hound make, to overtake the hare?

91. A grocer would mix the following kinds of sugar, viz. at 10 cents, 13 cents, and 16 cents per lb. What quantity of each must he take, to make a mixture worth 12 cents per lb.?

92. A grocer has 43 gallons of wine worth $1.75 a gallon, which he wishes to mix with another kind worth $1.40 a gallon, in such proportion that the mixture may

be worth $1.60 a gallon. How many gallons at $1.40 must he use?

93. Three merchants, A, B, and C, freight a ship with wine. A puts on board 500 tons, B 340 tons, and C 94 tons; and in a storm they are obliged to cast 150 tons overboard. What loss does each sustain?

94. A and B hired a pasture for 37 dollars. A put in 3 horses for 4 months, and B 5 horses for 3 months. What ought each to pay?

95. A family of 10 persons took a large house for $\frac{1}{2}$ of a year, for which they agreed to pay 500 dollars for that time. At the end of 14 weeks, they took in 4 new lodgers; and after 3 weeks, 4 more; and so on at the end of every 3 weeks, during the term, they took in 4 more. How much rent must one of each class pay?

96. A boy bought 12 apples and 6 pears for 17 cents, and then, at the same rate, 3 apples and 12 pears for 20 cents. What was the price of an apple, and of a pear?

97. A certain square pavement contains 48841 square stones, all of the same size. How many stones constitute the length of one side of the pavement?

98. A certain field lies in the form of a right-angled triangle; the sides containing the right angle are, one 48 rods, the other 20 rods in length. What is the length of the other side? How many acres in the field?

99. The following lots of sugar, from Havana, were sold in Boston on account of owners in Cuba, at $12\frac{3}{8}$ cents per lb. Required the amount of sales for each owner, allowing draft 4 lb. per box, and tare 15 per cent.

A's sugar, 21 boxes, weighing 10794 lb. gross.
B's sugar, 70 boxes, weighing 35980 lb. gross.
C's sugar, 84 boxes, weighing 43176 lb. gross.
D's sugar, 105 boxes, weighing 53970 lb. gross.

100. How much money on interest at the rate of 6 per cent. a year, from February 16th 1835, will be sufficient to meet a custom-house bond of $1464.45, which will become due on 10th of January, 1836?

101. How many shingles will cover the roof of a house, which is 40 feet in length, and has 30 feet rafters, supposing each shingle to be 4 inches wide, and each course to be 6 inches?

102. A merchant sold a piece of cloth for $40, and by so doing, lost 10 per cent. For how much should he have sold it, to have gained 15 per cent.?

103. A merchant received on consignment, three parcels of hops, viz. 450 lb. from Allen, 890 lb. from Brooks, and 510 lb. from Chase, Allen's hops were found on inspection to be $33\frac{1}{3}$ per cent. better than the others, but it was necessary to sell them together, at 12 cts. a pound. How much must each owner be credited?

104. Three parcels of beef, of 60 barrels each, were received at Baltimore, from Boston, marked, W, X, Y. The lot marked W was found to be .50 per cent. better than the others. The whole was sold together at 10 dollars a barrel. How must the sale be adjusted between the owners of the beef?

105. If iron worth $4, per cwt. cash, is sold for $4.50, on a credit of 8 months, what credit should be allowed on wine worth in cash $224 per pipe, but sold at $242, to make the percentage equal to that on the iron?

106. The number of terms in an equidifferent series is 11, the last term is 32, and the sum of the terms is 187. Find the first term, and the common difference.

107. A merchant has three notes, due to him as follows; one of $300, due in 2 months; one of $250, due in 5 months; and one of $180, due 3 months ago, with interest; the whole of which he now receives. What sum is received on the three notes, allowing money to be worth 6 per cent. a year?

108. A lady has two silver cups, and only one cover. The first cup weighs 12 oz. If the first cup be covered, it will weigh twice as much as the second; but if the second cup be covered, it will weigh three times as much as the first. What is the weight of the cover, and of the second cup?

109. Gray of Baltimore remits to Degrand in Boston, for sale, a set of exchange on London, the proceeds of which to be invested in certain merchandise for Gray's account. On selling the bill at 10 per cent. advance, D received $8600. How many pounds sterling was the bill drawn for, and how much is D to lay out for G, re-

serving to himself ½ of 1 per cent. on the sale of the bill, and 2½ per cent. commission on the investment?

110. The greatest term in a series of continual proportionals is 10, the ratio 1½, and the number of terms 12. What is the sum of the series?

111. What is the area of a circle, the diameter of which is 200 feet?

112. What sum of money must be put on interest, at the rate of 6 per cent. a year, to gain $27.83 in 11¾ months?

113. A person found two sums of money; ¼ of the first added to ⅓ of the second was $120. The two sums together were $400. What was each sum?

114. What number is that, whose cube root is equal to the square root of 361?

115. If a family of 9 persons spend $305. in 4 months, how many dollars would maintain them 8 months, if 5 persons more were added to the family?

116. Bought 5 hhds. of wine at 1 dollar per gallon, cash; having kept it 3 months and 23 days, I sold it at $1.20 per gallon, on a credit of 5 months; 16 gallons having leaked out while in my possession. What was my cash gain?

117. A grocer having sugars at $12, $10, and $8 per cwt. would make a mixture of 30 cwt. worth $9 per cwt. What quantity of each must he take?

118. What sum of money on interest at 6 per cent. a year, will amount to $1295.19, in 13 months 6 days?

119. How much must be paid for the transportation of 7261½ lb. 60 miles, at the rate of $10 for the transportation of 2000 lb. 47 miles?

120. A farmer sold 17 bushels of rye, and 13 bushels of wheat for $31.55. The wheat, at 35 cents a bushel more than the rye. What was each per bushel?

121. A man bought apples at 5 cents a dozen, half of which he exchanged for pears, at the rate of 8 apples for 5 pears; he then sold all his apples and pears at a cent apiece, and thus gained 19 cents. How many apples did he buy, and how much did they cost?

122. The sides of two square pieces of ground are as

3 to 5, and the sum of their superficial contents is 30600 square feet. What is the length of a side of each piece?

123. If 96 boards, 15 feet 6 inches long and 14 inches wide, will floor a place, how many will it take if the boards are only 11 feet 4 inches long and 9 inches wide?

124. A certain club spent at a supper, 43 dollars 56 cents, and the expense of each was as many cents as there were persons in the company. What did each pay?

125. If 20 feet of iron railing weigh 1000 lb. when the bars are $1\frac{1}{4}$ inch square, what will 50 feet come to, at $9\frac{3}{4}$ cents per lb. when the bars are $\frac{7}{8}$ of an inch square?

126. A stationer sold quills at $\$1.83\frac{1}{3}$ per thousand, and gained $\frac{1}{4}$ of the first cost; but quills growing scarce, he raised the price to $\$2.04$ per thousand. What did he gain per cent. by the last sale?

127. A merchant purchased goods to the amount of 3472 dollars, which he sold at a loss of $12\frac{1}{2}$ per cent. and invested the proceeds of the sale in other goods, which he sold at a profit of 13 per cent. Did he gain or lose by these transactions, and how much?

128. A house completely finished, has cost the owner $\$12894$; it is 4 stories high, and the ground floor is divided into two shops, one of which is let at $\$225$, the other at $\$200$ a year; the three upper stories are let for $\$450$ a year; the annual expense for repairs is $\$36.89$. What per cent. does the house pay?

129. A merchant in Boston received from New Orleans a bill at 30 days sight; he allowed 1 per cent. discount for present payment, and received $\$2530.44$. What sum was the bill drawn for; and what was the discount?

130. A merchant sold a parcel of coffee at 15 cents per lb. and lost 10 per cent.; soon after he sold another parcel, to the amount of $\$525$. and gained 40 per cent. How many pounds were there in the last parcel; and at what price per lb. was it sold?

131. G received from H 760 lb. of rough tallow to try out, at 60 cents per 100 lb. clear, and was to take his pay in rough tallow at 8 cents per lb.; G returned 615 lb. clear, and H paid him the balance due to him in rough tallow. Allowing 18 per cent. for waste, what was the balance due to G?

132. A merchant received on consignment 3 lots of hops, viz. 810 lb. from Allen; 720 lb. from Bond; and 1872 lb. from Cook. On inspection, Bond's hops were found to be $12\frac{1}{2}$ per cent. better than Allen's; and Cook's 25 per cent. better than Bond's. A sale of the 3 lots together was effected at 10 cents per lb. What was the just share of the amount for each?

133. A and B hired a coach in Boston to go 50 miles, for $25. with liberty to take in two more when they pleased. After riding 15 miles they took in C, who wished to go the remainder of the journey out, and return with them to Boston. On their return, at the distance of 25 miles from Boston, they admitted D for the remainder of the journey. You are required to settle the coach hire equitably between them.

134. Suppose a rope 7.1365 rods long, to have one end attached to a horse's head, and the other end fastened to a stake, in the centre of a field; how much land will the horse be allowed to graze upon?

135. A cistern is to be constructed, in the form of a cylinder, to hold 850 gallons. If the diameter of the end be made 6 feet, what must be the length of the side?

136. Suppose the propelling wheels of a locomotive engine to be 3 feet 4 inches in diameter, and to make 390 revolutions in a minute; what distance will the engine move forward in one hour?

137. If 12 oxen eat up $3\frac{1}{2}$ acres of grass in 4 weeks, and 21 oxen eat up 10 acres in 9 weeks, how many oxen will eat up 24 acres in 18 weeks; the grass being at first equal on every acre, and growing uniformly?

THE END.

INDEX.